Experiences in Math for Young Children

DEDICATION

To Our Dear Friend Ada Dawson Stephens

Experiences in Math for Young Children

COPYRIGHT © 1978
BY DELMAR PUBLISHERS INC.

LIBRARY OF CONGRESS CATALOG CARD NUMBER: 77-80039
ISBN: 0-8273-1660-7

Printed in the United States of America
Published simultaneously in Canada
by Nelson Canada,
A Division of International Thomson Limited

Rosalind Charlesworth

Deanna J. Radeloff

Jeanne M. Machado — Consulting Editor
Elinor Gunnerson — Early Childhood Education Series Editor

DELMAR PUBLISHERS INC.

Preface

EXPERIENCES IN MATH FOR YOUNG CHILDREN is designed for use by students in training and teachers in service in Early Childhood Education. To the student, it introduces the excitement and extensiveness of math experiences in programs for young children. For the teacher in the field, it presents an organized, sequential approach to developing a math curriculum.

Activities are presented in a developmental sequence designed to teach the concepts and skills needed in early childhood. An assessment approach is stressed in order to have an individualized program in which each child is presented with tasks that he can accomplish successfully before moving on to the next level.

A further emphasis is placed on three types of child learning: naturalistic, informal, and structured. Much of math learning can take place through the child's natural exploratory activities if the environment is designed to promote such activity. The adult can reinforce and enrich this naturalistic learning by careful introduction of information and structured experiences.

Math has been largely ignored as a curriculum area in early childhood. The types of concepts included are often referred to as "pre-math" apparently under the assumption that math learning only begins with addition and subtraction in the primary grades. This text is designed to erase that myth and bring to the attention of early educators the value and necessity of math experiences for young children.

Consistently throughout the text, *he* is used to refer to the student and *she* to the teacher except when specific behavioral examples are presented. This practice is for reasons of clarity only.

Both authors are currently Assistant Professors in the Early Childhood Education Program at Bowling Green State University, Ohio. Dr. Charlesworth holds a joint appointment in Educational Psychology in the Department of Educational Foundations and Inquiry and Child and Family Studies in the Department of Home Economics. Mrs. Radeloff's appointment is full-time in Child and Family Studies.

Mrs. Radeloff was instrumental in developing the Early Childhood Education major and minor at Bowling Green State University. Both authors have taken leadership in the implementation of this program. They are now developing a graduate level program in Early Childhood Education. Sections of this edition of *Experiences in Math for Young Children* were classroom tested at Bowling Green.

Dr. Charlesworth's career in Early Childhood Education has included experiences with both normal and atypical children in laboratory schools, public schools, and day care and research in social and cognitive development. She also taught courses in Early Education and Child Development at other universities before joining the faculty at Bowling Green.

Mrs. Radeloff spent several years teaching Vocational Home Economics and has coordinated a cooperative work experience program for senior high school students enrolled in a child care assistant program. She has been active in Child Development and Child Care Services Development for the Vocational Department of Home Economics in the State of Ohio. Mrs. Radeloff has had practical experience teaching young children and has been both a Head Start and day care consultant.

Other texts in the Delmar Early Childhood Education Series are

Creative Activities for Young Children, 3/e — Mayesky, Neuman, and Wlodkowski
Teaching Young Children — Beatrice Martin
Early Childhood Experiences in Language Arts, 3/e — Jeanne Machado
Administration of Schools for Young Children — Phyllis Click
Early Childhood: Development and Education — Jeanne Mack
Home and Community Influences on Young Children — Karen VanderVen
Experiences in Music for Young Children — Mary Carolyn Weller Pugmire
Early Childhood Education in the Home — Elinor Massoglia
Experiences in Science for Young Children — Donald Neuman
A Practical Guide to Solving Preschool Behavior Problems — Eva Essa
Understanding Child Development — Rosalind Charlesworth
Beginnings and Beyond: Foundations in Early Childhood Education — Gordon & Brown
Seeing Young Children: A Guide to Observation & Recording of Behavior — Warren R. Bentzen
Health, Safety, and Nutrition for the Young Child — Marotz, Rush and Cross
Early Childhood Practicum Guide — J. Machado and H. Meyer

A current catalog including prices of all Delmar educational publications is available upon request. Please write to:

Catalog Department
Delmar Publishers Inc.
2 Computer Drive, West
Albany, New York 12212

Contents

Appendices

Section 1 Math Development in Young Children

unit 1 how math develops

OBJECTIVES

After studying this unit, the student should be able to

- Define math development
- Identify examples of young children developing math ideas and skills
- Label examples of Piaget's developmental stages of thought
- Identify conserving and nonconserving behavior and state why conservation is an important developmental task

Math includes many ideas and skills which help people organize their world. The ideas and skills which young children learn and use include matching, counting, classifying, comparing, ordering, and measuring. They also learn about shape, space, number, and number symbols. As children grow and develop physically, socially, and mentally, their math ideas and skills grow and develop. *Development* means changes due to growth and experience over time. Development does not follow the same timetable for each child. It is a series or sequence of steps which each child reaches one at a time. Children may be weeks, months, or even a year or two apart in reaching certain stages and still be within the normal range of development.

Math growth and development begin in infancy. The baby explores the world with his senses. He looks, touches, smells, hears, and tastes. The child is born curious. He wants to know all about his environment. The baby begins to learn ideas of size, weight, shape, time, and space. As he looks about, he senses his relative smallness. He grasps things and finds that some fit his tiny hand and others do not. The infant learns about weight when items of the same size cannot always be lifted. He learns about shape. Some things stay where he puts them while others roll away. He learns time sequence. When he wakes up, he feels wet and hungry. He cries. Mom comes. He is changed and then fed. Next he plays, gets tired, and goes to bed to sleep. As the infant begins to move, he develops an idea of space. He is placed in a crib, in a playpen, or on the floor in the center of the livingroom. As the baby first looks and then

Fig. 1-1 The infant learns about the world through his senses.

1

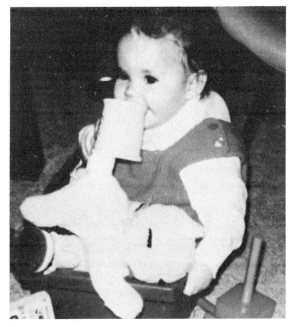

Fig. 1-2 The toddler learns about size, shape, and space.

moves, he discovers space. Some spaces are big. Some spaces are small.

As the child learns to crawl, to stand, and to walk, he is free to discover more on his own and learns to think for himself. He holds and examines more things. He goes over, under, and in large objects and discovers his size relative to them. The toddler sorts things. He puts them in piles — of the same color, the same size, the same shape, or with the same use. The young child pours sand and water into containers of different sizes. He piles blocks into tall structures and sees them fall and become small parts again. He buys food at a play store and pays with play money. As the child cooks imaginary food, he measures imaginary flour, salt, and milk. He sets the table in his play kitchen, putting one of everything at each place just as is done at home. The free exploring and experimentation of the first two years is the opportunity for the development of muscle coordination and the sense of taste, smell, sight, and hearing. The child needs these skills as a basis for future learning.

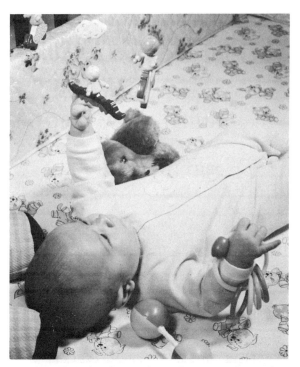

Fig. 1-3 The infant learns about distances as he reaches and grasps.

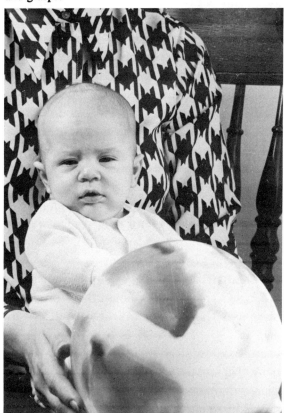

Fig. 1-4 The infant learns the shape idea of roundness.

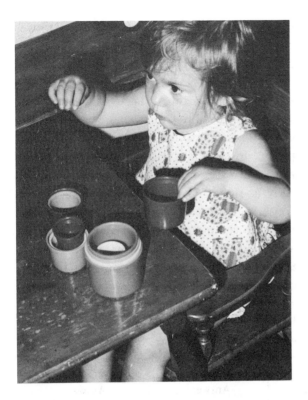

Fig. 1-5 **Sorting and matching are important toddler activities.**

Fig. 1-6 **The young child learns about measurement.**

PIAGETIAN STAGES OF MATHEMATICAL THOUGHT

Jean Piaget, a developmental psychologist, has contributed greatly to an understanding of the development of children. One area of Piaget's work is mathematical thought. Piaget has studied the development of thought through four stages of growth. Those who work with young children are concerned with the first two stages and the beginning of the third stage.

The first stage that Piaget identified, the *sensorimotor period* (from birth to about two years of age), has been described in the first part of this unit. It is the time when the child begins to learn. He makes use of all his senses — touch, taste, sight, hearing, and smell. He also makes use of his growing motor abilities — to grasp, to crawl, to stand, and to walk. The child in the first stage is an explorer and needs to have opportunities to use his sensory and motor abilities to learn basic skills and ideas.

The second stage is called the *preoperational period* and extends from about two years until around seven years of age. During this time the child begins to develop ideas and thoughts that are like those of adults. Language develops as the child talks more and

Fig. 1-7 **The young child learns that there is one buttonhole for each button.**

more. He learns more and more words and applies them to the ideas which began to develop during the first stage. He can use math words such as big and small, light and heavy, square and round, late and early, long and short, and so on.

An important characteristic of the pre-operational child is that he believes only what he sees. When materials are changed in form or arrangement in space, he may see them as changed in amount, also. From the adult point of view, he is easily fooled. For instance, if the same amount of drink is put in a tall thin glass as in a short fat glass, he will say there is more in the tall glass "because it is taller." If clay is changed from a ball shape to a snake shape, he will say there is less clay "because it is thinner." If a pile of coins is placed together, he will say there are fewer coins than if they are spread out. The child cannot hold the first picture of the coins in his mind when the material is changed.

During the third stage, called *concrete operations* (usually from ages seven to eleven), the child is able to remember or store the first picture in his mind. He is then said to be able to *conserve*. A child may begin to conserve as early as five or as late as seven. Being able to conserve is a good indicator that the child is ready to learn formal math such as addition and subtraction. This is very important for the teacher to know. Section 2 of this text discusses the basic skills and ideas the child seems to need in order to be able to conserve. Piaget's final stage is *formal operations* (ages eleven through adulthood).

Original	Physical Change	Question	Non-Conserving Answer	Conserving Answer
Same amount of drink. 		Is there still the same amount of drink?	No, there is more in the tall glass.	Yes, you just put the drink in different size glasses.
Same amount of clay. 	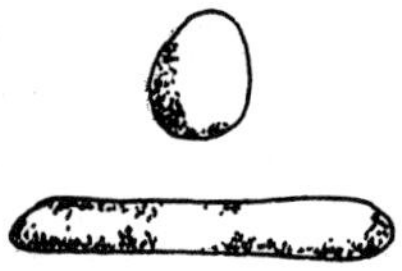	Is there still the same amount of clay?	No, there is more clay in the snake because it is longer.	Yes, you just rolled it out into a different shape.
Same amount of pennies. 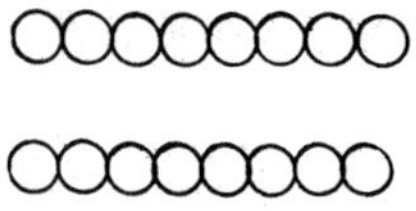	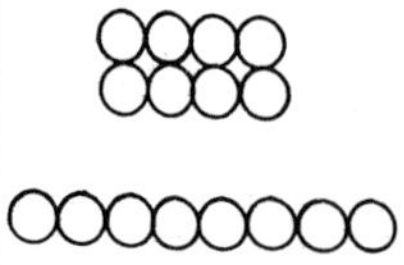	Are there still the same number of pennies?	No, there are more in the bottom row because it is longer.	Yes, you just moved the pennies closer together (points to top row).

Fig. 1-8 Physical Changes in Conversation Tasks

| Period | Skills and Ideas | | |
	Section II	Section III	Section IV
Sensorimotor (Birth to age two)	Matching Shape Space	Ordering	
Preoperations (Two to seven years)	Sets and Classifying Number and Counting Comparing Counting Parts and Wholes Language	Measuring: Weight, Length Temperature, Volume, Time and Sequence Graphical Representations	Number Symbols Sets and Symbols Formal Math: Such as addition, sub- traction, and so on.
Concrete Operations (Ages seven to eleven)			
Formal Operations (Age eleven on)	↓	↓	↓

Fig. 1-9 The Development of Math Skills and Ideas

THREE GROUPS OF MATH IDEAS

This book divides math ideas into three groups as can be seen in figure 1-9. The first group (section 2) includes skills and ideas which start to develop during the sensorimotor period and grow through the preoperational period. These include activities such as matching, counting, classifying, and comparing. Also included is an introduction to shape, space, parts and wholes, and the language of math. These skills are basic to conservation. The second group (section 3) lists applications of the first group. It includes ordering, measurement, pictorial representation, and practical life activities. The third group (section 4) includes number symbols and matching symbols with amounts. Also included are activ-

ities for the children who can conserve. As can be seen in figure 1-9, math development is not a quick — or a one-time — lesson. It starts in infancy and continues through all of early childhood and, of course, beyond.

SUMMARY

Development of math ideas and skills starts in infancy and grows through stages. The exploratory activities of the infant and toddler are the basis of later success. The use of the senses and muscles teaches the child about the world. Jean Piaget has developed a description of the stages of thought. He has studied the task of conservation which must be mastered before the child can be successful with formal math (such as addition and subtraction).

SUGGESTED ACTIVITIES

• Observe three different children at home or at school. One should be 6-18 months, one 18 months-2 1/2 years, and the third 3-5 years of age. Record everything each child does which you think indicates he is learning or has learned a math idea or skill. What differences did you observe among the three children?

• Interview three different mothers of children age 2-5 1/2 years. Ask them what kinds of math activities their children do. How do the mothers' views of math compare with what you have learned in this unit?

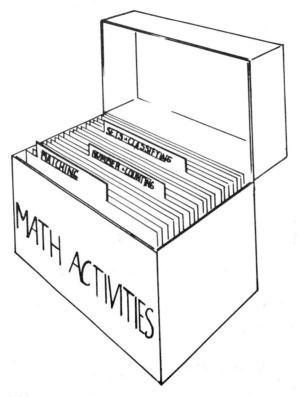

Fig. 1-10 Start a Math Activity File now so you can keep it up to date.

- Interview two or three young children. Present the problems illustrated in figure 1-8. Tape record their answers. Listen to the tape and describe what you learned. How did the children's answers differ? How were they similar?

- You should begin to record on 5 1/2″ x 8″ file cards each math activity that you learn about. Buy a pack of cards, some dividers, and a file box. Label your dividers with the titles of units 4 through 23. Figure 1-10 illustrates how your file should look.

REVIEW

A. Define the term *math development*.

B. Match each of the listed names of math ideas and skills with a correct description of child behavior from those which follow.

Math ideas and skills:

1. money
2. size
3. sorting

4. space
5. time

Description of child behavior:

a. The child examines objects of different sizes or compares himself with another person or thing.

b. The child goes through a series of regular events: arrives at school, has free play, eats a snack, paints or hears a story, toilets, washes, eats lunch, takes a nap, has a snack, plays outdoors, goes home.

c. The child discovers he can crawl under the table but not under the couch.

d. Two children are playing store and one exchanges some pieces of green paper for some food.

e. A child is putting blocks in different piles by color and size.

C. Decide which of the following descriptions describes a child in the sensorimotor (SM), preoperational (P), or concrete operational (CO) Piagetian stages.

1. Mary watches as her teacher makes two balls of clay of the same size. The teacher then rolls one ball into a snake shape and asks, "Mary, do both balls still have the same amount or does one ball have more clay?" Mary laughs, "They are still the same amount. You just rolled that one out into a snake."

2. Michael shakes his rattle and then puts it in his mouth and tries to suck on it.

3. John's mother shows him two groups of pennies. One group is spread out and one group is stacked up. Each group contains ten pennies. "Which bunch of pennies would you like to have, John?" John looks carefully and then says, "I'll take these because there are more," as he picks up the pennies that are spread out.

D. In review question C, which child, Mary or John, is a conserver? How do you know? Why is it important to know that a child is or is not a conserver?

unit 2 how math is learned

OBJECTIVES

After studying this unit, the student should be able to

- List and define the three types of learning experiences described in the unit
- Recognize examples of each of the three types of learning experiences
- State possible responses to specific opportunities for the child to learn math

There are three types of learning experiences for the young child. These three types may be called the naturalistic, the informal, and the structured. Math experiences occur in all three types.

NATURALISTIC LEARNING EXPERIENCES

Naturalistic experiences are those which are initiated by the child as he goes about his daily activities. Some of these were described in unit one as the main way of learning for the sensorimotor child. As the child learns to use his senses and his muscles, he learns math ideas (also called *concepts*) such as shape, size, time, and amount.

The adult's role is to provide an interesting and rich environment. That is, there should be many things for the child to look at, touch, taste, smell, and hear. The adult should observe the child's activity and note how it is progressing and then respond with a glance, a nod, a smile, or a word of praise to encourage the child. The child needs to know when he is doing the appropriate thing.

Some examples of naturalistic experiences are listed:

- Kurt hands Dad two pennies saying, "Here's your two dollars!"

- Tamara takes a spoon from the drawer — "This is big." Mom says, "Yes."
- Roger is eating orange segments. "I got three." (Holds up three fingers.)
- Nancy says, "Big girls are up to here," as she stands straight and points to her chin.

INFORMAL LEARNING EXPERIENCES

Informal experiences are initiated by the adult but are not preplanned. They occur when the adult's experience and/or intuition indicates it is time to act. This might happen for various reasons; such as, the child makes a mistake; the child is having difficulty; the child

Fig. 2-1 The child learns measurement in a naturalistic activity.

Fig. 2-2 "One, two, three.": An Example of Informal Learning

is on the right track in solving a problem but needs some cues. An unexpected opportunity to learn appears by chance. Some examples of the preceding reasons are given:

- "I'm six years old," says three-year-old Kate while holding up three fingers. Dad says, "Let's count those fingers. One, two, three fingers. You are three years old."

- Bob is setting the table. He gets frustrated because he does not seem to have enough cups. "Let's check," says his teacher. "There is one placemat for each chair. Let's see if there is one cup on each mat." They move around the table checking. They come to a mat with two cups. "Two cups," says Bob. "Hurrah!" says his teacher.

- With arms outstretched at various distances, Tim asks, "Is this big? Is this big?" Mr. Brown says, "What do you think? What *is* 'this' big?" Tim looks at the distance between his hands with his arms stretched to the fullest. "This is a big person." He puts his hands about eighteen inches apart. "This is a baby."

He places his thumb and index finger about half an inch apart. "This is a blackberry." Mr. Brown watches with a big smile on his face.

- Juanita has a bag of cookies. Mrs. Ramirez asks, "Do you have enough for everyone?" Juanita replies, "I don't know." Mrs. R. asks, "How can you find out?" Juanita says, "I don't know." Mrs. R. tells her, "I'll help you. We'll count them."

STRUCTURED LEARNING EXPERIENCES

Structured experiences are preplanned activities. They may be done with individuals or with groups. They may be done at a specific time or at an opportune time. The following are examples of types of structured activities:

- **With an individual at a specific time:** "Maria, I have some blocks here for you to count. How many in this pile?"

- **With a group at a specific time:** The teacher is seated with six children in a

Fig. 2-3 A structured learning situation: "Put them in order from smallest to largest."

semicircle in front of her. "Today we are going to talk about things that are large and small." She holds up a grapefruit and an orange. "Find the one that is large."

- **With an individual at an opportune time:** Mr. Flores knows that Tanya needs help in learning the names of shapes. He notices that she is playing with some cards that have pictures of shapes. He sits next to her and says, "Those are interesting. Can you find the cards with circles on them?"

- **With a group at an opportune time:** Mrs. Raymond has been working with the children on the ideas of light and heavy. She notices four children out on the playground trying to lift a large plank to make a ramp onto the packing box. "Is that board too light for you?" she asks with a twinkle in her eye.

SUMMARY

Three types of learning experiences have been described and defined. The teacher and parent learn through practice how to make the best use of naturalistic, informal, and structured experiences so that the child has a balance of free exploration and specific planned activities.

SUGGESTED ACTIVITIES

- Observe in an early childhood center. Keep a record of learning experiences which would be called naturalistic, those which would be called informal, and those which would be called structured. Which ones would you feel are math experiences?

- Observe in an early childhood center. Note times when you feel opportunities for naturalistic and informal learning are missed.

Fig. 2-4 Informal learning takes place with self-correcting materials.

Fig. 2-5 To solve the problem, the child must think hard.

Fig. 2-6 The child who watches also learns.

REVIEW

A.   List the three types of learning experiences and write your own definition or description of each.

B.   Indicate whether the examples which follow are naturalistic, informal, or structured.

1.   "Mama, I'll cut this donut in half and give you part." "Good idea," says Mom.

2.   Eighteen-month-old Brad has lined up four small dishes and is putting a toy dog in each one.

3.   Teacher and four children are sitting at a table. Each child has a pile of colored chips in front of him. "Line up three red chips." Children follow directions. "Put a blue chip next to each red chip."

4.   "I have three cookies," says Leroy. Teacher notices that Leroy has four cookies. "Let's count those cookies, Leroy, just to be sure."

5.   Three children are pouring water. They have many sizes and shapes of containers. "I need a bigger bowl, please," says one to another.

C.   Tell how you would react in the following situations. Would you respond with a naturalistic, an informal, or a structured learning experience?

1.   Richard and Diana are playing house. They are setting the table for dinner. They carefully place each place setting in front of each chair.

2.   Most of the children in the class seem to have trouble telling squares from rectangles.

3. Sam says, "I have more crayons than you have, George." "No, you don't." "Yes, I do!"

4. Pete is trying to put a round peg in a square hole. He is beginning to look upset.

5. The children need some help in understanding and using time words such as *yesterday, today,* and *tomorrow* and *early* and *late.*

   - "Hi, Diana. You are here *early* today."
   - "Hurry, everyone, or we will be *late* for lunch."
   - *"Yesterday* I told you we would have a surprise *today."*

unit 3 how math is taught

OBJECTIVES

After studying this unit, the student should be able to

- List in order and define the six steps in choosing math objectives and activities
- Identify definitions of assessment and evaluation
- Discuss the advantages of using the six-step method
- Identify examples of each of the six steps
- Describe the choices the teacher must make after evaluation
- Evaluate whether a teacher uses the six steps

A teacher must know her students well in order to help them learn to their full capabilities. She must choose objectives and activities with care so that the child starts from the place where he is and moves as fast and as far as he can. The steps to be followed to plan math experiences are ones which can be used to teach any subject. Six questions must be answered:

- Where is the child now? **Assess**
- What should he learn next? **Choose objectives**
- What should the child do in order to accomplish these objectives? **Plan experiences**
- Which materials should be used to carry through the plan? **Select materials**
- Do the plan and the materials fit? **Teach** (do the planned experiences with the child)
- Has the child learned what was taught (reached objectives)? **Evaluate**

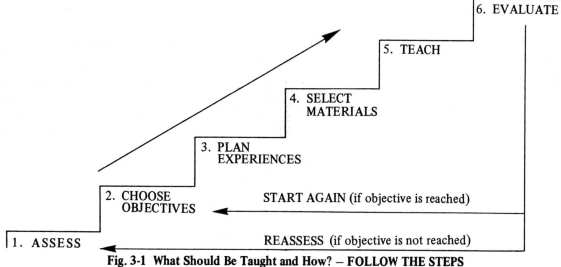

Fig. 3-1 What Should Be Taught and How? — FOLLOW THE STEPS

Fig. 3-2 Assessment, in an interview, of size concept: "Show me the one that is bigger."

ASSESSING

Each child should be individually assessed. There are two main ways to assess. The child can be given an individual interview using specific tasks. The child can be observed during his regular activities. The child is assessed to find out what he knows and what he can do before a math experience is presented. (Assessment is the topic of unit 4.)

Specific Task Assessment

The following are examples of some specific tasks that can be given to a child:

- Present the child with a pile of ten counters (buttons, coins, poker chips, or other small things) and say, "Count these for me."
- Show the child two groups of chips; a group of three and a group of six. Ask, "Which group has more chips?"
- Show the child five cardboard dolls, each one a half inch taller than the next. Say: "Which is the tallest?" "Which is the smallest?" "Line them up from the smallest to the tallest."
- Show the child cards with one shape drawn on each card: triangle, circle, rectangle, diamond, and square. Say, "Find the square." or "Tell me the name of each shape."

Assessment By Observation

Some examples of observations that could be made as the child plays are listed.

- Does the one-year-old show an interest in experimenting by pouring things in and out of containers of different sizes?
- Does the two-year-old spend time sorting objects and lining them up in rows?
- Does the three-year-old show an interest in understanding size, age, and time by asking how big he is, how old he is, and when will. . .questions?
- Does the four-year-old set the table correctly? Does he ask for help to write numerals, and does he use them in his play activities?
- Can the five-year-old divide a bag of candy so that each of his friends receives an equal share?

Through observation the teacher can find out if the child can apply math ideas and skills to real life problems and activities. By keeping a

Fig. 3-3 Assessment is done through observation of children using materials.

record of these observations, the teacher builds up a more complete picture of the child's strengths and weaknesses.

CHOOSING OBJECTIVES

Once the teacher knows the child's present level, she can choose the next objectives; that is, she can decide what the child should learn next. For instance, look at the first task example in the previous section. Suppose the child is four years old and counts fifty things correctly. The objective for this child would be different from that for another four-year-old child who counts three things correctly but not four or more. The first child probably does not need any more help with counting. The second child does. The next objective might be for the second child to be able to count four things correctly.

Suppose the teacher observes that a two-year-old spends very little time sorting objects and lining them up in rows. The teacher knows that this is a vital activity for a child of this age. She must choose an objective. The objective could be that the child be able to spend at least five minutes sorting and organizing small toys. Once the teacher chooses the goal, she needs to decide how she will help the child reach it.

PLANNING EXPERIENCES

Planning involves making decisions as to the best way for each child to arrive at the chosen objective. Will naturalistic and informal experiences be enough, or should there be some structured activities also? Will the child learn best on a one-to-one basis with an adult? In a small group? On his own? Or should he work with other children? Once these questions are answered, materials can be chosen. Sections 3, 4, and 5 of this book tell how to plan each kind of math experience that young children need.

SELECTING MATERIALS

Two things must be considered in choosing materials. First, there are some general features of good materials. They should be made to last and made so children can safely use them independently. They should also be useful for more than one kind of activity.

Within the six-step procedure, it is important to select materials that help the child acquire the chosen objectives. Materials should teach the particular idea or skill for which they are to be used.

In each unit of this book there are examples of materials. Section 5 contains lists and descriptions of many that are excellent.

TEACHING

Once the teacher has decided what the child will learn next and how she is going to help him learn it, she then proceeds with the teaching step. Teaching is the actual doing of the planned experiences using the selected materials. If she has gone through the first four steps with care, the experience should go smoothly. The child is likely to be interested and to proceed through the activities because they fit his level of development and way of learning. He may learn a new math idea or skill or expand one already learned.

The time involved in the teaching step may be a few minutes or several weeks, depending on the particular idea or skill being taught and the age and capability of the child. For instance, time sequence is often taught by marking special events and times on a calendar (see unit 15). The children practice using terms such as yesterday, today, and tomorrow. They learn to name the days of the week in order and learn the number of days in a week. This is a very complex experience — involving many ideas and skills. It is one which can be started with prekindergarten children and continued for many years in different ways.

Fig. 3-4 Teaching: "Feel the shape and tell me what it is."

Some children will move ahead fast and others slowly. One may learn the idea that there are seven days in every week the first time he is told. Another may need several months of work before he learns it. Some children need a lot of structured repetition while others learn from informal experiences. Teaching math the developmental way involves planning a flexible, individualized program.

Even with careful planning, an activity may not work out the first time. The teacher then looks at what happened and asks the following questions:

- Was the child interested?
- Was the task too easy or too hard?
- Did the child understand what he was asked to do?
- Were the materials right for the task? Were they interesting?
- Is further assessment needed?
- Was the teacher enthusiastic?
- Was it just a "bad" day for the child?

The teacher may decide to try the activity again the same way, or she may change the instructions and/or materials. In some cases she may have to assess the child again to be sure she has used an activity at his developmental level.

EVALUATING

The sixth step for the teacher is evaluation. What has the child learned? What does he know and what can he do after the math experiences have been presented? The assessment questions are asked again. If the child has reached the objective, a new one can be chosen. The steps of planning, choosing materials, teaching, and evaluating are gone through again. If the child has not reached the objective, the same activities can be continued or a new method may be tried. For example, a teacher wants a five-year-old to count out the correct number of objects for each of the number symbols from zero to ten. She tries many kinds of objects for the child to count and many kinds of containers in which to place the things he counts, but the child is just not interested. Finally she gives him small banks made from baby food jars and real pennies. The child finds these materials are exciting and goes on to learn the task quickly and with enthusiasm.

Evaluation may be done using formal, structured questions and tasks and specific observations as will be presented in unit 4. Informal questions and observations of naturalistic experiences can be used for evaluation also. For example, when a child sets the table in the wrong way, it can be seen without formal questioning that he has not learned from instruction. He needs some help. Maybe organizing and placing a whole table setting is more than he can do now. Can he place one item at each place? Does he need to go back to working with a smaller number (such as a table for two or three)? Does he need to work with simpler materials which have more structure (such as pegs in a pegboard)? To look at these more specific skills, the teacher would then return to the assessment step. At this point she would assess not only the child but also the types of experiences and materials she has been using. Sometimes assessment leads the teacher to the right objective but the experience and/or materials chosen are not (as in the example given) the ones that fit the child.

Frequent and careful evaluation helps both teacher and child avoid frustration. An adult must never take it for granted that any one plan or any one material is the best choice for a specific child. The adult must keep checking to be sure the child is learning what the experience was planned to teach him.

Fig. 3-5 Evaluation: "Do both rows have the same number of pennies?"

SUMMARY

This unit has described six steps that provide a guide for what to teach and how to teach it. Following these steps cuts down on guesswork. The steps are (1) assess, (2) choose objectives, (3) plan experiences, (4) choose materials, (5) teach, and (6) evaluate.

SUGGESTED ACTIVITIES

- Interview two early childhood teachers. Ask them to tell you what kinds of math experiences they include in their programs. Find out how they decide what to teach, to whom, and which materials to use. Go through their answers later and try to evaluate whether they use any or all of the steps described in this unit.

- Go to the library and look through recent issues of professional publications for teachers of young children. For your Activities File make a card for each article you find that gives ideas for using assessment and evaluation to help in choosing objectives, planning, and choosing materials. Summarize on each card the ideas which you feel will be helpful to you.

- Spend a morning in an early childhood center. Note all the math experiences. Go through your notes and evaluate what you observed. What steps did you see? Did you see any incidents where you felt that the teacher needed to evaluate her own teaching or the child's learning? Why?

REVIEW

A. Place the following list of steps for choosing objectives and activities in the correct order.

1. Evaluate
2. Plan experiences
3. Select materials
4. Teach
5. Assess
6. Choose objectives

B. Write the names of the steps that go with the definitions below.

1. Finding out what the child knows and can do *before* you present a math experience.

2. Finding out what the child knows and can do *after* you present a math experience.

3. Deciding what the child should learn next.

4. Making decisions as to the best way for each child to arrive at the chosen objectives.

5. Doing the planned experiences using the selected materials.

6. Selecting materials that help the child acquire the chosen objectives.

C. Read the following descriptions and label them with the correct step name.

1. The teacher goes to her Activities File and looks in the section on materials. She takes out the cards labeled *Measurement,* looks through them, and takes out two.

2. The teacher looks over the results of the assessment tasks she has just given to Jimmie. She sees he has difficulty with *more* and *less.* He can tell when the differences are large but not when they are small. There has to be at least four more in one group for him to be able to label the groups correctly. She tries to decide on the next step for Jimmie.

3. The teacher is seated at a table with Cindy. On the table are many different objects. "Cindy, we've been learning about how different things belong together. You put the things from this pile into smaller piles of things which belong together." After Cindy finishes, the teacher asks, "Tell me why these belong together." She points to each pile in turn.

4. It is the beginning of the year. Before planning her math program Mrs. Ramirez questions each child individually to find out exactly which skills and ideas he has and at what stage he is.

5. Mr. Brown is having the children learn how to use measuring cups and spoons in preparation for making a chocolate cake. They are using the cups with sand and water in order to discover the relationships between the different size cups.

6. The next objective for Kate is to be able to correctly organize a three-part-sequence story. Mrs. Raymond goes to her Activity File and looks under *Time and Sequence.*

D. Discuss the advantages of following the steps presented in this unit.

E. After evaluation what choices does the teacher have?

F. Read in the following paragraph about Miss Collins' way of choosing objectives and activities. Analyze and evaluate her approach in terms of what you have learned in this unit.

> Miss Collins believes that all children are about the same unless they have an extreme handicap. Her math program is the same from year to year. She assumes that all the children coming into her class need to learn the same things with the same materials. She feels that she does a fine job and that when her children leave, they are all ready for kindergarten work — although she has never questioned them at the end of the year to find out.

unit 4
how to assess the child's developmental level

OBJECTIVES

After studying this unit, the student should be able to

- Explain how to find the child's level of development in math
- Make a developmental assessment task file
- Recognize tasks to be used at each developmental level
- Do developmental math assessment of young children

The child's developmental level in math is found by seeing which math tasks he is able to do. The first question in teaching is "Where is the child now?" To find the answer to this question the teacher assesses. The teacher gives the child tasks to solve (such as those described in unit 3). She observes what the child does as he solves the problems and records the answers he gives. This information is used to guide the next steps in teaching. The long term practical objective for the young child is to help him to someday have the skills and ideas needed to do formal math (such as to add and subtract). Following the methods and sequence in this text helps reach this goal and at the same time achieves some further objectives:

- Builds a positive feeling in the child toward math
- Builds confidence in the child that he can do math tasks
- Builds a questioning attitude in response to his curiosity regarding math problems

ASSESSMENT METHODS

Observation and interview are assessment methods the teacher uses to find out the child's level of development. Examples of both of these methods were included in unit 3. More are given in this unit.

Observation Assessment

Observation is used to find out how the child uses his math ideas and skills in his daily experiences. The teacher has in mind the math ideas and skills that the children could be using. Whenever she sees an idea expressed or a skill used, she writes down the incident and places it in the child's record folder. This helps her plan the child's future math experiences.

Throughout this book, suggestions are made for behaviors which should be observed.

Fig. 4-1 The teacher learns the child's concept of her body in space.

The following are examples of behaviors as the teacher would write them down for the child's folder:

- Brad (eighteen months old) dumped all the shape blocks on the rug. He picked out all the circles and stacked them up. Shows he can sort and organize.

- Cindy (four years old) carefully set the table for lunch all by herself. She remembered everything. Understands one-to-one correspondence.

- Chris (three years old) and George (five years old) stood back to back and asked Cindy to check who was taller. Good cooperation — first time Chris has shown an interest in comparing heights.

- Mary (five years old), working on her own, put the right number of sticks in juice cans marked with the number symbols zero through twenty. She is ready for something more challenging.

Interview Assessment

The individual interview is used to find out specific information in a direct way. The teacher can present a task to the child and observe and record the way the child works on the task and the solution he arrives at for the problem presented by the task. The rightness and wrongness of the answer is not as important as how the child arrives at the answer. Often a child starts out on the right track but gets off somewhere in the middle of the problem. For example, Kate (age three) is asked to match four saucers with four cups. This is an example of matching skills. She does this task easily. Next she is asked to match five cups with six saucers, "Here are some cups and saucers. Find out if there is a cup for every saucer." She puts a cup on each saucer. Left with an extra saucer, she places it under one of the pairs. She smiles happily. By observing the whole task, the teacher can

Fig. 4-2 The teacher looks and listens as the children compare heights.

see that Kate does not like groups with an unequal amount of things. This is normal for a preoperational three-year-old. She finds a way to do away with the problem by putting two saucers under one cup. She understands the idea of matching one to one but cannot have things out of balance. Only by observing the whole task can the teacher see the reason for what appears to be a "wrong" answer to the task.

For another example, Tim who is just four-and-one-half is given the following task. First he is shown cards with the number symbols zero to six. He is asked to name each number symbol and does so correctly. He is then asked to place the correct number of chips by each number symbol. Tim's responses tell the teacher that he recognizes and can name number symbols but that he cannot yet match the symbols with the right number of chips. He can recognize groups up to four but does not yet have the idea of groups of more than four. He tried to count out five chips and six chips but lost track after four. As will be seen in unit 6, Tim's behavior is normal for a four-year old.

If Kate's and Tim's answers were observed only at the end point and recorded as right or wrong, the true level of the child's development would be missed. Only the individual interview offers this opportunity for the teacher to observe a child solve a problem from start to finish with no distractions or interruptions.

An important factor in the one-to-one interview is that it must be done in an accepting manner by the adult. She must value and accept the child's answers whether they are right or wrong from the adult point of view. If possible, the interview should be done in a quiet place where there are no other things which might take the child's attention off the task. The adult should be warm, pleasant, and calm. Let the child know that he is doing well with smiles, words ("Good," "Fine," "You're a good worker," "Keep trying hard"), and gestures (nod of approval, pat on the shoulder).

If persons other than one of the teachers do the assessment interviews, the teacher should be sure that they spend time with the children before the interviews. Advise a person doing an interview to sit on a low chair or on the floor next to where the children are playing. Children are usually curious when they see a new person. One may ask, "Who are you? Why are you here?" The children can be told, "I am Ms. X. Someday I am going to give each of you a turn to do some special work with me. It will be a surprise. Today I want to see what you do in school and learn your names." If the interviewer pays attention to the children and shows an interest in them and their activities, they will feel comfortable and free to do their best when the day comes for their assessment interview.

If the teacher does the assessment herself, she also should stress the special nature of the activity for her and each child, "I'm going to spend some time today doing some special work with each of you. Everyone will get a turn."

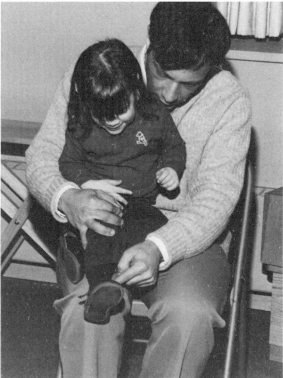

Fig. 4-4 **A parent volunteer meets the children before doing any interviews.**

Fig. 4-3 **The child enjoys the individual interview.**

ASSESSMENT TASK FILE

Each child and each group of children is different. The teacher needs to have on hand questions to fit each age and stage she might meet in individual young children. She also needs to add new tasks as she discovers more about children and their development. A card file of assessment tasks should be set up. Such a file has three advantages:

- The teacher has a personal involvement in creating her own assessment tasks and is more likely to use them, understand them, and value them.
- The file card format makes it easy to add new tasks and revise or remove old ones.
- There is room for the teacher to use her own creativity to add new questions and make materials.

Use the tasks in the Appendix to begin the file. Other tasks can be developed as the student proceeds through the units in this book and through her future career with young children. Directions for each task can be put on five-by-eight-inch plain white file cards. Most of the tasks will require the use of concrete materials and/or pictures. Concrete materials can be items found around the home and center. Pictures can be purchased or cut from magazines and readiness type workbooks and glued on cards.

The basic materials needed are listed: 5" x 8" file card box, 5" x 8" unlined file cards, 5" x 8" file dividers, black pen, set of colored markers, ruler, scissors, glue, clear Contac® or laminating material, and preschool/kindergarten readiness workbooks with artwork (such as the Golden Readiness Workbooks published by Western Publishing Company).

In the Appendix each assessment task is set up as it would be on a five-by-eight-inch card. Note that on each card what the adult says to the child is always printed in CAPITAL LETTERS so the instructions can be found and read easily. The tasks are set up develop-

mentally from the sensorimotor (birth to age two) level through the preoperational (ages two to seven) level. Ages are given to serve as a guide for choice of the first tasks to present to each child.

Each child is at his own level. If the first tasks are too hard, the interviewer should start at a lower level. If the first tasks are quite easy for the child, the interviewer should start at a higher level. Sections 3, 4, and 5 tell how to proceed when this first step in teaching is finished. Figure 4-5, page 24, is a sample recording sheet that could be used to keep track of each child's progress. The name and number of the task is put in the first column, the date in the next, and the results in the third. The second set of columns for each task is for reassessment.

THE ASSESSMENT TASKS

The tasks included in the Appendix represent the ideas and skills which must be demonstrated by the young child before he is taught formal math. Most of the tasks require an individual interview with the child. Some tasks are observational and require recording of activities during playtime. The infant tasks and observations assess the development of the child's growing sensory and motor skills. As was discussed in the first unit, these sensory and motor skills are basic to all later learning.

The assessment tasks are divided into eight developmental levels. *Levels One and Two* are tasks for the child in the sensorimotor stage. *Levels Three through Five* include tasks of increasing difficulty for the prekindergarten child. The *Level Six* tasks are those things which the child can usually do when he enters kindergarten between the ages of five and six. This is the level he is growing toward during his prekindergarten years. Some children will be able to accomplish all these tasks by age five; others not

Child's Name _____ Birthdate _____

School _____ Teacher _____

Name of person doing assessment _____

Task	Date	Comments	Date	Comments

Comments:

Fig. 4-5 Recording Sheet for Developmental Tasks

Fig. 4-6 **Even the toddler works hard on the interview tasks.**

Fig. 4-7 **Free Sort, Three-year-old: "Which ones belong together?"**

until six or over. *Level Seven* summarizes the math words which are usually a part of the child's natural speech by age six. *Level Eight* is included as an assessment for advanced prekindergarteners and for children in centers that have a kindergarten program. The child about to enter first grade should be able to accomplish the tasks at *Level Six* and *Level Eight*. He should also be using most of the math words correctly.

EXAMPLE OF AN INDIVIDUAL INTERVIEW

The following is a part of the *Level Five* assessment interview as given to Bob (four and one-half years old). A corner of the storage room has been made into an assessment center. Mrs. Ramirez comes in with Bob, "You sit there and I'll sit here, Bob. We have some important things to do." They both sit down at a low table and Mrs. Ramirez begins.

Fig. 4-8 **Free Sort, Four-year-old: A Challenging Task.**

Fig. 4-9 **The five-year-old tries to find out if there are the same number of players on each team.**

Mrs. Ramirez:	Bob's Response:
HOW OLD ARE YOU?	"I'm four." (He holds up four fingers.)
COUNT TO TEN FOR ME, BOB. (Mrs. Ramirez nods her head up and down)	"One, two, three, four, five, six, seven, eight, nine, ten,. . .I can go some more. Eleven, twelve, thirteen, twenty!"
HERE ARE SOME BLOCKS. HOW MANY ARE THERE? (She puts out ten blocks.)	(He points, saying) "One, two, three, four, five, six, seven, eight, nine, ten, eleven, twelve." (He points to some more than once.)
GOOD, BOB. NOW COUNT THESE. (Five blocks)	(He counts, pushing each one he counts to the left) "One, two, three, four, five."
She puts the blocks out of sight and brings up five plastic horses and five plastic cowboys.	
FIND OUT IF EACH COWBOY HAS A HORSE.	(Bob looks over the horses and cowboys. He lines up the horses in a row and then puts a cowboy on each.) "Yes, there are enough."
FINE, BOB. (She puts the cowboys and horses away. She takes out some inch cube blocks. She puts out two piles of blocks: five yellow and two orange.	
DOES ONE GROUP HAVE MORE?	"Yes." He points to the yellow.
GOOD. She puts out four blue and three green.	
DOES ONE GROUP HAVE LESS?	He points to the green blocks.
WELL DONE.	
She takes out five cutouts of bears of five different sizes. FIND THE BIGGEST BEAR.	"Here it is." (He picks the right one.)
FIND THE SMALLEST BEAR. PUT ALL THE BEARS IN A ROW FROM BIGGEST TO SMALLEST.	(He points to the smallest.)
(Mrs. Ramirez smiles.)	(Bobby works slowly and carefully.) "All done." (Two of the middle bears are reversed.)
GOOD FOR YOU, BOB. YOU'RE A HARD WORKER.	

An interview does not have to include any special number of tasks. For the preoperational child, the teacher can begin with matching and proceed through the ideas and skills one at a time so that each interview can be quite short if necessary.

If the person doing the interviews has time for longer sessions they should plan as follows:

- Fifteen to twenty minutes for two-year-olds

- Thirty minutes with three-year-olds

- Forty-five minutes with four-year-olds

- Up to an hour with five-year-olds

Fig. 4-10 "Put these in a row from the skinniest to the fattest."

Fig. 4-11 "This is a triangle."

SUMMARY

There are two ways children can be assessed to find their developmental level. They can be observed and they can be interviewed. Observation is most useful when looking at how children use math in their everyday activities. The interview with one child at a time gives the teacher an opportunity to look at very specific ideas and skills.

Guidelines are given for doing an interview. There is a summary of the eight levels of developmental tasks which are included in the Appendix. A sample of part of an interview shows how the exchange between interviewer and child might progress.

SUGGESTED ACTIVITIES

- Interview the kindergarten supervisor (or if your system starts at grade one, the primary supervisor) in your local public school system. Find out what kinds of math knowledge and skills the children are expected to have when they enter school (kindergarten or grade one). Compare their list with the tasks in *Level Six* (and *Seven* if the children start at grade one): Are they the same? What differences are there? Would you make some changes to be sure the children who leave your center are ready for your local schools?
- Contact the office which supervises school psychological services in your local school district. Find out if there is preschool testing (sometimes called *screening*) for children entering the system for the first time. If there is a preschool testing program, find out what kinds of

tasks are used. Compare the math questions used by the school district with the ones suggested in the Appendix of this text. Are the questions about the same? Do they cover the same topics? Would the child who succeeds at *Levels Six, Seven* (and *Eight* for grade one entrance) be ready for the kindergarten (or first grade) programs of your system? What changes might you have to make in the tasks from the text?

- Find at least two children at two different age levels that you could have permission to assess in math. Make up the assessment cards and gather the materials needed. Try out the tasks with the children. Discuss with the other students in your class any problems you discovered. As a group, work out improvements in the method and list suggestions for making the assessment run smoothly.

- Based on the results recorded for the assessments done in the preceding activity, write out some teaching objectives for the children in the areas where they had the most difficulty. Look ahead in the text and find the units which tell how to teach in the areas assessed. Pick out some activities that fit your objectives.

- Invite a kindergarten and/or a first grade teacher to visit your class and describe how they assess and evaluate the levels of their students in math.

- Go to the library. Find five articles in periodicals which discuss assessment and/or evaluation of young children. List the main ideas about which you read. What did you learn that will help you in the future? Was math mentioned as a separate area or were the assessments for broader areas?

REVIEW

A. Explain how to find a child's level of development in math. Why is it important to make this assessment as the first step in teaching?

B. Read incidents 1-4 which follow. What is being done wrong in each situation? What should be done?

1. Mr. Brown is interviewing a child in the teachers' lounge. Other teachers are coming in and going out. There is a lot of talking among the adults. The child keeps looking away from the assessment materials to see who is in the room.

2. Ms. Collins has a volunteer parent come in to do some of her assessment tasks. She tells the parent, "Just go ahead and take the children in order from the attendance list. The faster the better."

3.	Mr. Flores is interviewing Jimmy. Mr. Flores puts a marble and a ping pong ball on the table. WHAT IS THIS? (Points to marble)	"A marble."
	WHAT IS THIS? (Points to ball)	"A ping pong ball."
	IS ONE HEAVIER THAN THE OTHER OR ARE THEY BOTH THE SAME?	"The ping pong ball is heavier because it is bigger." (Jimmy looks closely at Mr. F.'s face which has a very serious expression.)
4.	Mrs. Raymond is interviewing Cindy. Mrs. R places a nickel, a dime, a penny, and a dollar bill on the table. TELL ME THE NAME OF EACH OF THESE.	
	OKAY, RIGHT.	"Penny." (Points to penny.)
	WRONG, YOU HAVE THOSE MIXED UP.	"Dime." (Points to nickel.) "Nickel." (Points to dime.)

C. Decide at which developmental level each of the following tasks would be placed. Use the tasks in the Appendix as a guide.

1. The child is shown a clock: WHAT IS THIS? WHAT DOES IT TELL US?

2. Between the child and the adult on the table are two rows of poker chips matched one to one. The adult says, NOW, WATCH WHAT I DO. She moves the chips in the row nearest her so they are touching each other in the middle of the row rather than being an inch apart.

3. The adult watches as the child places plastic beads in a coffee can and then dumps them out again.

4. The adult notes that the child has just begun to line his blocks and toy cars up in rows.

5. The child is shown drawings of a triangle, a square, and a circle. He is asked to point to each as it is named.

6. The child holds a rattle for two or three seconds.

29

D. The teacher is doing some math assessments. What should she do in each of the following situations?

    1. She has a group of infants ranging in age from three months to twelve months. Another teacher tells her that these children are too young to learn math. What could the first teacher say to show that math learning begins in infancy?

    2. She is starting out with a group of children ages four to five years. She knows nothing about their level of ideas and skills in math.

    3. The teacher is giving the *Level Three* tasks to Pete who is two and one-half years old. She finds that he can give his age correctly and hold up two fingers. He can count five objects and can state which of two dolls is the big one. He can follow directions and put a block in, on, under, next to, and over a cup. The teacher has noticed that he makes trains when playing with blocks and puts people on each block to ride the train.

    4. Mary is four years old. When asked how old she is, she holds up three fingers and says, "I'm six." When asked to count to ten she says, "One, two, six, four, ten." The teacher gives her ten blocks to count. Mary points in a disorganized way as she says, "One, two, five, eight, six!"

    5. The teacher gives five-year-old Bobby two groups of chips: ten red and ten blue. FIND OUT IF THERE IS THE SAME NUMBER IN EACH GROUP. Without hesitation Bobby counts each group, "They are the same — ten in each group. I can prove it to you." He stacks a red on each blue and quickly counts the ten stacks correctly. "That was easy."

E. The best review of this unit is to do the third and fourth activities in the suggested activities section. Make an assessment kit and find out what it is like to use the materials with children.

Section 2 The Basics of Math

unit 5 matching

OBJECTIVES

After studying this unit, the student should be able to

- Define matching
- Identify naturalistic, informal, and structured matching activities
- Describe five ways to vary matching activities
- Assess and evaluate a child's matching skills

Matching, or one-to-one correspondence, is the most basic component of number. It is the understanding that one group has the same number of things as another. For example, each child has a cookie; each foot has a shoe; each man wears a hat.

ASSESSMENT

The teacher should note the child's free play and routine activities. She looks for naturalistic matching. For example, when the child plays train, he may line up a row of chairs so there is one for each child passenger. When he puts his mittens on, he shows that he knows that there should be one for each hand; when painting, he checks to be sure he has each paintbrush in the matching color of paint. Tasks for formal assessment are given in Appendix A. The following are examples:

Sample Assessment Task   **Preoperational Ages 2-3**

Matching (unit 5): One: one

Play activities: Does the child match items one to one such as putting a small peg doll in each of several margarine containers or on top of each of several blocks which have been lined up in a row?

Sample Assessment Task   **Preoperational Ages 5-6**

Matching (unit 5): One:one

Present the child with two groups of different shaped objects (pennies and cubes, for example), each group with ten items. FIND OUT IF THERE IS THE SAME AMOUNT (NUMBER) IN EACH BUNCH (GROUP, SET). Does the child arrange each group so as to match one to one the objects from each group?

NATURALISTIC ACTIVITIES

Matching activities develop from the infant's early sensorimotor activity. He finds out that he can hold one thing in each hand, but he can put only one object at a time in his mouth. The toddler discovers that five peg dolls will fit one each in the five holes in his toy bus. Quickly he learns that one person fits on each chair, one shoe goes on each foot, and so on. The two-year-old spends a great deal of his playtime matching things. He lines up containers such as margarine cups, dishes, or boxes and puts a small toy animal in each one. He pretends to set the table for lunch. First he sets one place for himself and one for his bear with a plate for each. Then he gives each place a spoon and a small cup and saucer.

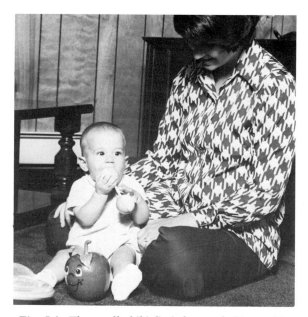

Fig. 5-1 The small child finds he can hold one thing in each hand, but can put only one object at a time in his mouth.

He plays with his large plastic shapes and discovers there is a rod that will fit through the hole in each shape.

INFORMAL ACTIVITIES

There are many opportunities for informal matching activities each day. There are many times when things must be passed out to a group: food items, scissors, crayons, paper, napkins, paper towels, or notes to go home. Each child should do as many of these things as possible.

Checking on whether everyone has accomplished a task or has what they need is another chance for matching informally. Does everyone have a chair to sit on? Does everyone have on two boots or two mittens? Does each person have on his coat? Does each person have a cup of milk or a sandwich? A child can check by matching: "Larry, find out if everyone has a pair of scissors, please."

One-to-one correspondence helps to solve difficulties. For instance, the children are washing rubber dolls in soap suds. Jeanie is crying, "Petey has two dolls and I don't have

Fig. 5-2 "Which rod goes with which shape?"

any." Mrs. Carter comes over. "Petey, more children want to play here now so each one can have only one baby to wash." One-to-one correspondence is often the basis for rules such as "Only one person on each swing at a time." "Only one piece of cake for each child today."

Other informal activities occur when children pick out materials made available

Fig. 5-3 One block is matched with each square.

during free play. These kinds of materials would include pegboards, felt shapes on a flannel board, bead and inch-cube block patterns, shape sorters, formboards, lotto games, and other commercial materials. Materials can also be made by the teacher to serve the same purposes. Most of the materials described in the next section can be made available for informal use by the children after they have been introduced in structured activities.

STRUCTURED ACTIVITIES

The extent and variety of matching activities is almost endless. There are five characteristics of matching activities:

- Perceptual characteristics

- Number of items to be matched

- Concreteness

- Physically joined or not physically joined

- Groups of the same or not the same number

The teacher can vary or change one or more of the five characteristics and can use different materials. In this way, more difficult tasks can be designed.

Perceptual qualities are very important in matching activities. The way the materials to be matched look is important in determining how hard it will be for the child to match them. Materials can differ a great deal on how much the same or how much different they look. Materials are easier to match if the groups are different. To match animals with cages or to find a spoon for each bowl is easier than making a match between two groups of blue chips. In choosing objects, the task can be made more difficult by picking out objects that look more the same.

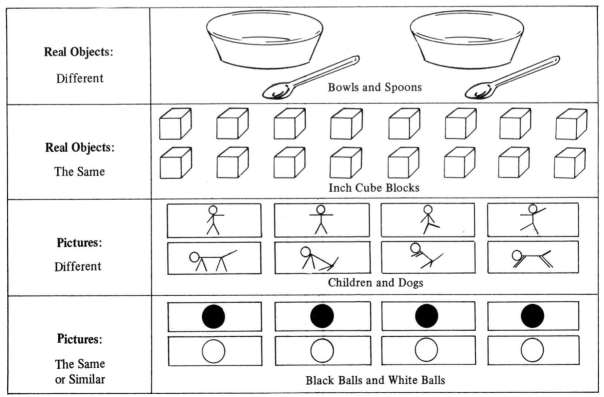

Fig. 5-4 Examples of Groups with Different Perceptual Difficulty Levels

The number of objects to be matched is important. The more objects in each group, the more difficult it is to match. Groups with less than five things are much easier than groups with five or more. In planning activities, start with small groups (less than five) and work up step by step to groups of nine. When the child is able to match groups of nine, this indicates he understands one to one correspondence.

How close to real the things to be matched are is called concreteness. The less concrete (or less real) the more difficult it is to match. The easiest and first matching tasks should use real things such as small toys and other familiar objects. The next level would be cut-out shapes such as circles and squares or less familiar, similar looking objects such as poker chips, inch cube blocks, or popsicle sticks. Next would come pictures of real objects and pictures of shapes. Last would come the least concrete: dots and symbols, figure 5-5.

It is easier to tell if a match is right if the objects are joined than if they are not joined. In beginning activities, the objects may be hooked together with a line or a string so the child can see that there is or is not a match. In figure 5-6, each foot is joined to a shoe and each animal to a cage. The hands and mittens and balls and boxes are not joined.

To match unequal groups is harder than to match equal groups. When the groups have the same number, the child can check to be sure he has used all the items. When one group has more, he does not have this clue, figure 5-7.

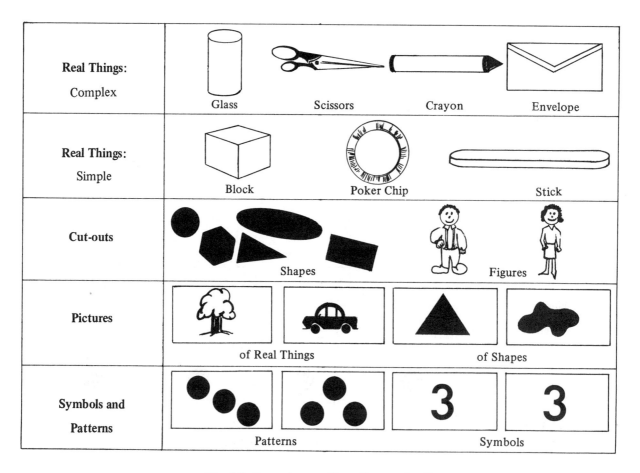

Fig. 5-5 Concreteness: How Close to Real?

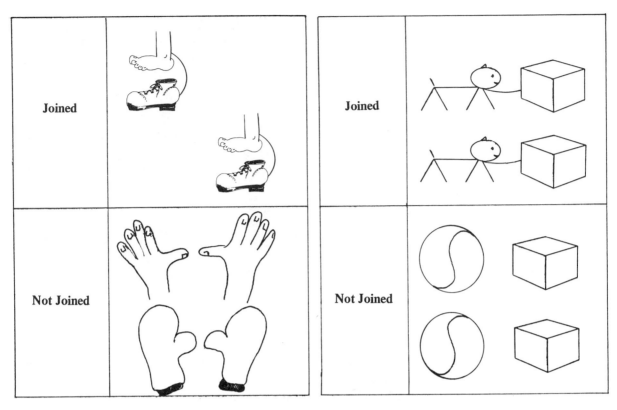

Fig. 5-6 Joined Groups and Not-joined Groups

Fig. 5-7 Matching Equal and Unequal Groups

The sample lessons that are given illustrate some basic types of matching activities. Note that they also show an increase in difficulty by varying the five characteristics just described. The lessons are shown as they would be put on cards for the Idea File, figure 5-9 through figure 5-14.

EVALUATION

Informal evaluation can be done by noticing each child's response during structured activities. Also observe each child during free play to see whether he can pass out toys or food to other children, giving one at a time. On the shelves in the housekeeping area, paper shapes of each item (such as dishes, cups, tableware, purses, etc.) may be placed on the shelf where the item belongs. Hang pots and pans on a pegboard with the shapes of each pot drawn on the board. Do the same for blocks and other materials. Notice which

Fig. 5-8 These children learn math and social cooperation.

children can put materials away by making the right match.

Using the same procedure as in the assessment tasks a formal check can be made. Once the child can match two groups of nine

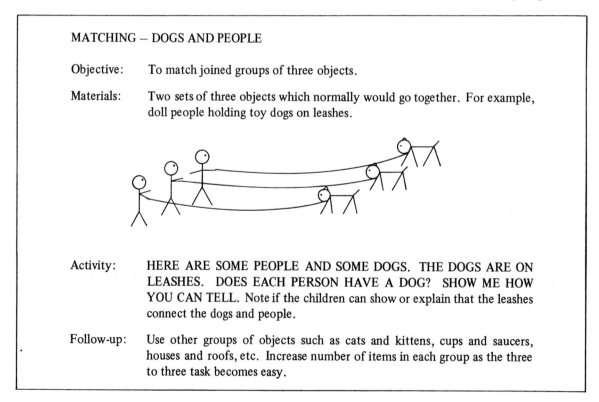

MATCHING – DOGS AND PEOPLE

Objective: To match joined groups of three objects.

Materials: Two sets of three objects which normally would go together. For example, doll people holding toy dogs on leashes.

Activity: HERE ARE SOME PEOPLE AND SOME DOGS. THE DOGS ARE ON LEASHES. DOES EACH PERSON HAVE A DOG? SHOW ME HOW YOU CAN TELL. Note if the children can show or explain that the leashes connect the dogs and people.

Follow-up: Use other groups of objects such as cats and kittens, cups and saucers, houses and roofs, etc. Increase number of items in each group as the three to three task becomes easy.

Fig. 5-9 Matching Activity Card – Dogs and People: Matching Objects that are Perceptually Different

MATCHING — THE THREE PIGS

Objective: To match joined groups of three items.

Materials: Two sets of cut-outs for the bulletin board. Three pieces of yarn. Make
three pig cut-outs and three house cut-outs: straw, sticks, and bricks. Put
them on bulletin board and connect each pig to his house with thick yarn:

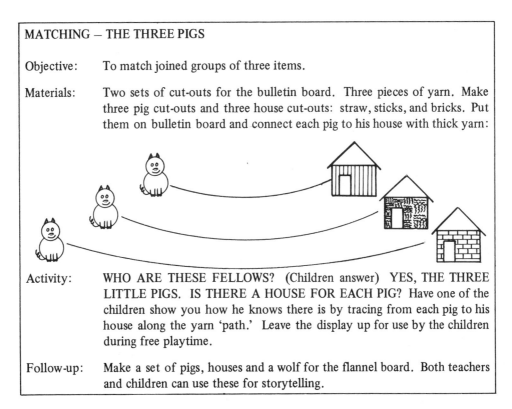

Activity: WHO ARE THESE FELLOWS? (Children answer) YES, THE THREE
LITTLE PIGS. IS THERE A HOUSE FOR EACH PIG? Have one of the
children show you how he knows there is by tracing from each pig to his
house along the yarn 'path.' Leave the display up for use by the children
during free playtime.

Follow-up: Make a set of pigs, houses and a wolf for the flannel board. Both teachers
and children can use these for storytelling.

Fig. 5-10 Matching Activity Card — The Three Little Pigs Matching Cut-Outs that are Perceptually Different

MATCHING — PENNIES FOR TOYS

Objective: To match groups of two and more objects.

Materials: Ten pennies and ten small toys (for example, a ball, a car, a truck, three
animals, three peg people, a crayon).

Activity: LET'S PRETEND WE ARE PLAYING STORE. HERE ARE SOME
PENNIES AND SOME TOYS. Show the child(ren) two toys. Place two
pennies near the toys. DO I HAVE ENOUGH PENNIES TO BUY THESE
TOYS IF EACH ONE COSTS ONE PENNY? SHOW ME HOW YOU CAN
FIND OUT.

Follow-up: Use more toys and more pennies as the children can match larger and
larger groups.

Fig. 5-11 Matching Activity Card — Pennies for Toys: Matching Real Objects

MATCHING – PICTURE MATCHING

Objective: To match groups of pictured things, animals, or people.

Materials: Make or purchase picture cards which show items familiar to young children. Each set should have two groups of ten. Pictures from catalogues, magazines, or readiness workbooks can be cut out, glued on cards, and covered with clear Contac or laminated. For example, pictures of ten children should be put on ten different cards. Pictures of ten toys could be put on ten other cards.

Activity: Present two people and two toys. DOES EACH CHILD HAVE A TOY? SHOW ME HOW YOU CAN FIND OUT. Increase the number of items in each group.

Follow-up: Make some more card sets. Fit them to current science or social studies units. For example, if the class is studying jobs, have pilot with plane, driver with bus, etc.

Fig. 5-12 Matching Activity Card – Picture Matching

MATCHING – SIMILAR OR IDENTICAL OBJECTS

Objective: To match two to ten similar and/or identical objects.

Materials: Twenty objects such as poker chips, inch cube blocks, coins, cardboard circles, etc. There may be ten of one color and ten of another or twenty of the same color (more difficulty perceptually).

Activity: Begin with two groups of two and increase the size of the groups as the children are able to match the small groups. HERE ARE TWO GROUPS (BUNCHES, SETS) OF CHIPS (BLOCKS, STICKS, PENNIES, ETC.). DO THEY HAVE THE SAME NUMBER OR DOES ONE GROUP HAVE MORE? SHOW ME HOW YOU KNOW. Have the children take turns using different sizes of groups and different objects.

Follow-up: Glue some objects to a piece of heavy cardboard or plywood. Set out a container of the same kinds of objects. Have this available for matching during free playtime.

Fig. 5-13 Matching Activity Card – Similar or Identical Objects

MATCHING – OBJECTS TO DOTS

Objective: To match 0–9 objects with 0–9 dots.

Materials: Ten frozen juice cans with dots (filled circles) painted a dark color on each
from zero to nine:

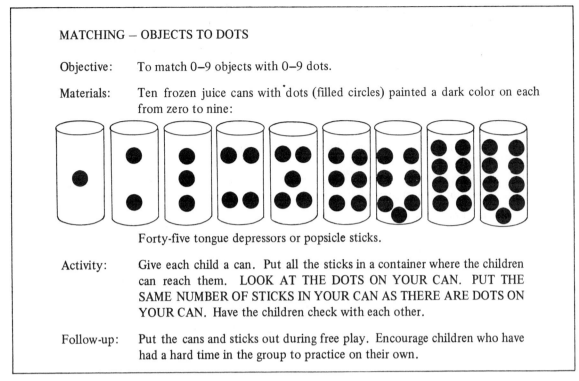

Forty-five tongue depressors or popsicle sticks.

Activity: Give each child a can. Put all the sticks in a container where the children
can reach them. LOOK AT THE DOTS ON YOUR CAN. PUT THE
SAME NUMBER OF STICKS IN YOUR CAN AS THERE ARE DOTS ON
YOUR CAN. Have the children check with each other.

Follow-up: Put the cans and sticks out during free play. Encourage children who have
had a hard time in the group to practice on their own.

Fig. 5-14 Matching Activity – Matching Objects to Dots

or more identical items, he does not need any more structured matching activities.

SUMMARY

The most basic number skill is matching or one to one correspondence. Starting in infancy the child learns about one to one relationships. The sensorimotor and early preoperations child spends much of his playtime in matching activities.

Many opportunities for informal matching are available during play and daily routines. Structured activities which can be presented to individuals and/or small groups can be put out for use during free play also.

Materials and activities can be varied in five different ways to make matching fun and interesting. Once the child can match two groups of nine objects on his own, he can be left to use his matching skills in solving everyday problems and as a basis for doing the activities in section 4.

SUGGESTED ACTIVITIES

- Give a two-year-old some small containers and some safe objects that will fit in the containers. Observe his play. Note any evidence of one to one matching.

- Select two or more of the structured activities from this unit. Prepare the materials needed and do the activities with two four-year-olds and two five-year-olds. Report the results to the class.

- Create a matching game. Try it out with a child. Show it to the class. Describe any improvements you would make next time.

- Copy the sample structured activities and place them in your Idea File. Add two more activities using different kinds of materials.
- Discuss in a small group what you would do if you had a five-year-old in school who could not do one to one matching of two groups of ten objects.

REVIEW

A. Define matching. Give at least one example.

B. Decide whether each of the following matching activities is naturalistic, informal, or structured.

    1. Mrs. Carter has some pictures. There are six dogs and six bones. "Petey, does each dog have a bone?"

    2. Kate lines up some square blocks in a row. Then she puts a wooden doll on each block.

    3. Cindy puts one shoe on each of her doll's feet.

    4. Mary carefully passes one cup of milk to each child.

    5. Mrs. Carter shows five-year-old George two groups of ten chips. "Find out if each group has the same number."

C. Describe five ways to vary the activities of matching things. Give examples.

D. Which of each of the following groups (a or b) would be harder to match?

    1. a. 4 blue chips to 4 white chips
        b. 10 blue chips to 10 white chips

    2. a. 5 shoes and 5 socks
        b. 5 blobs and 5 squares

    3. a. groups which are separated
        b. groups which can have each pair joined with a string

    4. a. two groups of six
        b. a group of five and a group of six

    5. a. cards with pictures of cups and saucers
        b. real cups and saucers

unit 6 number and counting

OBJECTIVES

After studying this unit, the student should be able to

- Define rote counting and rational counting and explain their relationship
- Identify examples of rote and rational counting
- Teach naturalistic, informal, and structured counting activities appropriate to each child's age and maturity level

Counting is a basic math skill that is too often overlooked as an instructional area for young children. Many teachers assume that counting is learned at home. However, this is often not the case. This is soon discovered as the teacher does her activities. Number is understanding the "oneness" of one, "twoness" of two, and so on. Zero to four are usually understood without counting. Some children can look at five and know it is five without having to count. To tell "how many" there are in groups of more than five, the child must count.

Counting includes two operations — rote counting and rational counting. *Rote* counting involves reciting the names of the numerals in order from memory. That is, the child who says, "One, two, three, four, five, six, seven, eight, nine, ten" has correctly counted in a rote manner from 1–10. *Rational* counting involves attaching each numeral name in order to a series of objects in a group. For example, the child has some pennies in his hand. He takes them out of his hand one at a time and places them on the table. Each time he places one on the table he says the next number name in sequence. "One," places first penny. "Two," places another penny. "Three," places another penny. He has successfully done rational counting of three objects. Rational counting is a higher level of matching, or one-to-one correspondence.

Rote counting is usually mastered ahead of rational counting. The examples above would be quite normal behavior for the same two and one-half to three year-old child. He might be able to rote count to ten or more but only rational count correctly groups of two or three. When given a group of more than three to count, he might perform as described in the following example. Six blocks are placed in front of a two and one-half year old, and he is asked, "How many

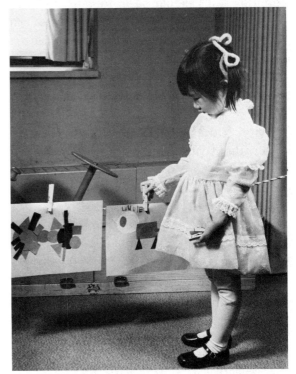

Fig. 6-1 "One clothespin; two clothespins."

blocks do you have? Using his pointer finger, he "counts" and points:

- "One, two, three, four, five, six, seven, eight" (pointing at some blocks more than once and some not at all)
- "One, two, four, six, three, ten" (pointing to each block only once but losing track of the correct order)

Counting things is much more complicated than reciting number names in order from memory. The child must coordinate eyes, hands, speech, and memory to correctly do rational counting. This is difficult for the two- or three-year-old because he is still in a period of rapid growth in coordination. He is also limited in his ability to stick to a task. The teacher should not push a child to count more things than he can do easily and with success. Most rational counting experiences should be naturalistic and informal.

By age four or five the rate of physical growth is slowing. Coordination of eyes, hands, and memory is maturing. Rational counting skills should begin to catch up with rote counting skills. Structured activities can be introduced. At the same time naturalistic and informal activities should continue.

ASSESSMENT

The adult should note the child's regular activity. Does he recognize groups of zero to four without counting: "Mary has no cookies"; "I have two cookies"; "John has four cookies." Does he use rational counting correctly when needed: "Here, Mr. Black, six blocks for you." (Mr. Black has seen him count them out on his own.) For formal assessment, see the tasks in the Appendix. The following are sample tasks:

Sample Assessment Task   Preoperational Ages 3-6
Number and Counting (unit 6): Rote Counting
COUNT FOR ME. COUNT AS FAR AS YOU CAN.
If the child hesitates or looks puzzled ask again. If he still does not count, Say, ONE, TWO, WHAT'S NEXT?

Sample Assessment Task   Preoperational Ages 3-6
Number and Counting (unit 6): Rational Counting
Place a pile of chips in front of the child (about ten for a three-year-old, twenty for a four-year-old, and thirty for a five-year-old). COUNT THESE FOR ME. HOW MANY CAN YOU COUNT?

NATURALISTIC ACTIVITIES

Number is an idea and counting is a skill which is used a great deal by young children in their everyday activities. Once these are in the child's thoughts and activity he will be observed often in number and counting activities. He practices rote counting often. He may run up to the teacher or parent saying, "I can count — one, two, three." He may be watching a TV program and hear "one, two, three, four...." He may then repeat "one, two, ..." At first he may play with the number names saying to himself, "one, two, five, four, eight, ..." in no special order. Listen carefully and note that gradually he gets more of the names in the right order.

Fig. 6-2 Rational counting involves coordination of memory, eyes, and hands.

Number appears often in the child's activities once he has the idea in mind. A child is eating crackers, "Look at all my crackers. I have two crackers." One and two are usually the first amounts used by children two and three years old. They may use one and two for quite a while before they go on to larger groups. Number names are used for an early form of division. That is, a child has three cookies which he divides with his friends, "One for you, one for you, and one for me." The child is looking at a picture book, "There is one daddy and two babies." The child wants another toy, "I want one more little car, Dad."

INFORMAL ACTIVITIES

When an adult recognizes that a child can learn from informal number and counting activities, she can help him. For example, the child is watching a children's television program. The teacher is sitting next to him. A voice from the TV rote counts by singing, "One, two, three, four, five, six." The teacher says to the child, "That's fun, let's count too." They then sing together. "One, two, three, four, five, six." Or, the teacher and children are waiting for the school bus to arrive. "Let's count as far as we can while we wait. One, two, three, . . ." Since rote counting is learned through frequent but short periods of practice, informal activities should be used most for teaching.

Everyday activities offer many opportunities for informal rational counting and number activities. For instance, the teacher is helping a child get dressed after his nap. "Find your shoes and socks. How many shoes do you have? How many socks? How many feet?" Some children are meeting at the door. The teacher says, "We are going to the store. There should be five of us. Let's count and be sure we are all here."

Fig. 6-3 Setting the table is a chance for informal counting.

Table setting offers many chances for rational counting. "Put six placemats on each table." "How many more forks do we need?" "Count out four napkins." Play activities also offer times for rational counting. "Mary, please give Tommy two trucks." A child is looking at his hands which are covered with fingerpaint: "How many red hands do you have, Joey?"

STRUCTURED ACTIVITIES

Rote counting is learned mostly through naturalistic and informal activities. However, there are short, fun things which can be used to help children learn the number names in the right order. There are many rhymes, songs, and fingerplays. Songs include those such as *This Old Man, Ten Little Indians,* and *The Twelve Days of Christmas.*

A favorite old rhyme is given:

> One, two, buckle your shoe.
> Three, four, shut the door.
> Five, six, pick up sticks.
> Seven, eight, shut the gate.
> Nine, ten, a big fat hen.

A finger play can be used:

Five Little Birdies

(Hold up five fingers. As each bird leaves fly your hand away and come back with one less finger standing up.)

Five little birdies sitting by the door
One flew away and then there were four.
Four little birdies sitting in a tree
One flew away and then there were three.
Three little birdies sitting just like you
One flew away and then there were two.
One little birdie sitting all alone
He flew away and then there were none.

More direct ways of practicing rote counting are also good. Clapping and counting at the same time teaches number order and gives practice in rhythm and coordination. With a group, everyone can count at the same time, "Let's count together. One, two, three, ..." Individual children can be asked, "Count as far as you can."

Groups which have zero to four items are special in the development of rational counting skills. The number of items in groups this size can be perceived without counting. For this reason these groups are easy for the child to understand. The child should have many experiences and activities with groups of size zero to four before he works with groups of five and more. With structured activities it is wise to start with groups of size two because, as mentioned before, there are so many which occur naturally. For example, the child has two eyes, two hands, two arms, and two legs. Two pieces of bread are used to make a sandwich, and bikes with two wheels are signs of being big. For this reason, activities using two are presented first in the following examples. The activities are set up so that they can be copied onto activity cards for the file.

Fig. 6-4 "One, two, three."

NUMBER: GROUPS OF TWO

Objective: To learn the idea of two

Materials: The children's bodies, the environment, a flannel board and/or a magnetic board, pairs of objects, pictures of pairs of objects

Activities:

1. Put several pairs of felt pieces (such as hearts, triangles, or bunnies, for example) on the flannel board (or magnets on the magnet board). Point to each group in turn: WHAT ARE THESE? HOW MANY ARE THERE?

2. Have the children check their bodies and the other children's bodies for groups of two.

3. Have the children, one at a time, find groups of two in the room.

4. Using rummy cards, other purchased picture cards, or cards you have made, make up sets of cards with identical pairs. Give each child a pack with several pairs mixed up. Have them sort the pack and find the groups of two.

5. Have a container with many objects. Have the children sort out as many groups of two as they can find.

Follow-up: Have the materials available during free play.

──────NUMBER: GROUPS OF THREE──────

Objective: To learn the idea of three

Materials: Flannel board and/or magnet board, objects, picture cards

Activities: Do the same types of activities using groups of three.

Follow-up: Have the materials available during free play.

──────NUMBER: GROUPS OF ONE──────

Objective: To learn the idea that one is a group

Materials: Flannel board and/or magnet board, objects, and picture cards

Activities: Do the same types of activities using groups of one as were done for groups of two and three.

Follow-up: Have the materials available during free play.

──────────NUMBER: ZERO──────────

Objective: To understand the idea that a group with nothing in it is called 'zero'

Materials: Flannel board, magnet board, and objects

Activities:

1. Show the children groups of things on the flannel board, magnet board and/or groups of objects. SEE ALL THESE THINGS? Give them a chance to look and respond. NOW I TAKE THEM AWAY. WHAT DO I HAVE NOW? They should respond with "nothing," "all gone," and/or "no more."

2. Put out a group of flannel pieces, magnet shapes, or objects of a size the children all know (such as one, two, three, or four). Keep taking one away. HOW MANY NOW? When none are left say: THIS AMOUNT IS CALLED ZERO. Repeat until they will answer "zero" on their own.

3. Play a Silly Game. Ask HOW MANY REAL LIVE TIGERS DO WE HAVE IN OUR ROOM? (continue with other things which obviously are not in the room).

Follow-up: Work on the concept informally. Ask questions: HOW MANY CHILDREN ARE HERE AFTER EVERYONE GOES HOME? (after snack if all the food has been eaten) HOW MANY COOKIES (CRACKERS, PRETZELS, etc.) DO YOU HAVE NOW?

After the children have the ideas of groups of zero, one, two and three then go on to four. Use the same kinds of activities. When they have four in mind then do activities using groups of all five amounts.

NUMBER: USING SETS ZERO ──────────THROUGH FOUR──────────

Objective: To understand groups of zero through four

Materials: Flannel board, magnet board, and/or objects

Activities:

1. Show the children several groups of objects of different amounts. Ask them to point to sets of one, two, three, and four.

2. Give the children a container of many objects. Have them find sets of one, two, three, and four.

3. Show the children containers of objects (pennies, buttons, etc.). Ask them to find the ones with groups of zero, one, two, three, and four.

4. Give each child four objects. Ask them to make as many different groups as they can.

5. Ask the children to find out HOW MANY ____ ARE IN THE ROOM? (Suggest things for which there are four or less.)

When the children have the idea of groups from zero to four they can then go on to groups larger than four. Some children are able to perceive five without counting just as they perceive zero through four without actually counting. Having learned the groups of four and less, children can be taught five by adding on one more to groups of four. When the children understand five as four with one more and six as five with one more and so on, then real rational counting can begin. That is, children can work with groups of objects where they can only find the number by actually counting each object. Before working with a child on counting groups of six or more, the adult must be sure the child can do the following:

- Recognize groups of zero to four without counting

- Rote count to six or more correctly and quickly

- Recognize that a group of five is a group of four with one more added

The following are activities for learning about groups larger than four.

NUMBER/RATIONAL COUNTING: INTRODUCING FIVE

Objective: To understand that five is four with one more item added

Materials: Flannel board, magnet board, and/or objects

Activities:

1. Show the children a group of four. HOW MANY IN THIS GROUP? Show the children a group of five. HOW MANY IN THIS GROUP? Note how many already have the idea of five. Tell them YES, THIS IS A GROUP OF FIVE. Have them make other groups with the same amount using the first group as a model.

2. Give each child five objects. Ask them to identify how many objects they have.

3. Give each child seven or eight objects. Ask them to make a group of five.

Follow-up: Have containers of easily counted and perceived objects always available for the children to explore. These would be items such as buttons, poker chips, Unifix blocks, and inch cubes.

NUMBER/RATIONAL COUNTING: GROUPS LARGER THAN FIVE

Objective: To be able to count groups of amounts greater than five

Materials: Flannel board and/or magnet board, objects for counting, pictures of groups on cards, items in the environment

Activities:

1. One step at a time present groups on the flannel board and magnet board and groups made up of objects such as buttons, chips, popsicle sticks, inch cube blocks, etc. Have the children take turns counting them — together and individually.

2. Present cards with groups of six or more, showing cats, dogs, houses, or similar figures. Ask the children as a group or individually to tell how many items are pictured on each card.

3. Give the children small containers with items to count.

4. Count things in the room. HOW MANY TABLES (CHAIRS, WINDOWS, DOORS, CHILDREN, TEACHERS)? Have the children count all at the same time and individually.

Follow-up: Have the materials available for use during free play. Watch for opportunities for informal activities.

NUMBER/RATIONAL COUNTING: FOLLOW-UP WITH STORIES

Objective: To work on rational counting

Materials: Stories which will reinforce the ideas of groups of numbers and rational counting skills: Some examples are *The Three Pigs, The Three Bears, The Three Billy Goats Gruff, Snow White and the Seven Dwarfs, Six Foolish Fishermen.*

Activities: As these stories are read to the younger children, take time to count the number of characters who are the same animal or same kind of person. Use felt cutouts of the characters for counting activities and for matching (as suggested in unit 5). Have older children dramatize the stories. Get them going with questions such as:

HOW MANY PEOPLE WILL WE NEED TO BE BEARS? HOW MANY PORRIDGE BOWLS, SPOONS, CHAIRS, BEDS) DO WE NEED? JOHN, YOU GET THE BOWLS AND SPOONS.

Follow-up: Have the books and other materials available for the children to use during free playtime.

NUMBER/RATIONAL COUNTING:
—FOLLOW-UP WITH COUNTING BOOKS—

Objective: To strengthen rational counting skills

Materials: Counting books (see list in Appendix)

Activities: Go through the books with one child or individual children. Discuss the pictures as a language development activity and count the items pictured on each page.

RATIONAL COUNTING: FOLLOW-UP
————WITH MATCHING————

Objective: To combine matching and counting

Materials: Flannel board and/or magnet board and counting objects

Activities: As the children work with the counting activities, have them check their sets which they say are the same number by using matching (one-to-one correspondence). See activities for unit 5.

Follow-up: Have materials available during free playtime.

EVALUATION

Informal evaluation can be done by noting the answers given by the children during teaching sessions. The teacher should also observe the children during free play and notice whether they apply what they have learned. When they choose to explore materials used in the structured lessons during the free play period, questions can be asked. For instance, Kate is at the flannel board arranging the felt shapes in rows. As her teacher goes by, she stops and asks, "How many bunnies do you have in that row?" Or, four children are playing house, and the teacher asks, "How many children in this family?" Formal evaluation can be done with one child by using tasks such as those in the Appendix.

SUMMARY

Number is understanding "oneness," "twoness," and so on. Counting may be rote or it may be rational. Rote counting is saying from memory the names of the numerals in the correct order. Rational counting is attaching the names in order to things in a group to find out how many items are in the group.

Rote counting is mastered before rational counting. Rational counting usually begins to catch up with rote counting after ages four or five. Number is learned at about the same time. Number amounts greater than four are not learned until the child learns rational counting.

Rote and rational counting may be learned through both naturalistic practice and informal and formally structured lessons.

SUGGESTED ACTIVITIES

- Present the following counting tasks to three or four children at different age levels between 2 1/2 and 6 1/2 years of age.
  A. COUNT FOR ME. SAY ALL THE NUMBERS YOU KNOW.
  B. Place a bunch (five for a two-year-old, ten for a three-year-old, fifteen for a four-year-old, etc.) of objects that do not roll (such as checkers, poker chips, popsicle sticks, inch-cube blocks) in front of a child: COUNT AS MANY AS YOU CAN (or HOW MANY *object name* DO I HAVE?)

Record the results and compare the answers received from the different age children.

- Go to the library. Find five number books (such as those listed in the Appendix). Go through each one. Write a description of each one. Tell whether you would buy it for your own school and how you would use it.

- Design a counting lesson of three or four activities. Try it out with individuals or small groups of children. Evaluate its success in a few sentences.

- Add five cards with rote counting activities and five cards with rational counting activities to your card file of activities.

- Collect small objects (such as buttons, toothpicks, etc.) and containers (such as empty pharmacists' pill bottles, egg cartons, etc.) and make your own counting kit.

- Visit a classroom. Write down a brief description of each example of a counting activity observed. Label each as to whether it involved rote or rational counting and whether it was naturalistic, informal, or structured teaching.

- View some children's educational television programs. Describe how counting is taught. Develop a plan for following up on what the child views in order to strengthen his learning.

REVIEW

A. Discuss each of the following.

1. *Rote* counting.

2. *Rational* counting.

B. Select the correct statements:

1. Rational counting develops before rote counting.

2. Rote counting is learned mainly through frequent, short periods of practice.

3. Most rote counting activities are naturalistic and informal.

4. Rote counting usually develops ahead of rational counting until about age four or five when rational counting starts to catch up.

5. Rational counting should be taught by having daily structured lessons beginning at age two.

6. When teaching rational counting, it is important to give the child groups of real objects to count.

C. Decide which of the following incidents would involve rote counting and which would involve rational counting.

 1. Mary says to her teacher, "I am going to get six napkins to put on the table."

 2. Johnny says, "I can count! Listen to me: one, two, three!"

 3. The children are singing, "This old man he played one... ."

 4. The teacher says to Joyce, "Please bring three more boxes of crayons."

 5. The teacher hands Peter a string of colored beads, "How many beads on the string, Peter?"

 6. "Let's count and clap — one (clap), two (clap). . . ."

D. What should the adult do in the following situations?

 1. Randy (age 3 1/2) looks up with a big smile and says, "One, two, three, four, five, six, seven, eight, nine, ten."

 2. Tanya (age 2 1/2) says, "I have two eyes, you have two eyes."

 3. Mary is five. She cannot count out eight forks for setting the table without making a mistake.

 4. Bobby is 4 1/2. He is counting a stack of five blocks: "One, two, four, eight, seven — seven blocks."

 5. The teacher is having lunch at a table with five three- and four-year-olds. She wants to have some informal counting experiences.

 6. Some of the children who are of kindergarten age seem to be having trouble with counting activities.

unit 7 sets and classifying

OBJECTIVES

After studying this unit, the student should be able to

- Describe features of sets and the activity of classifying.
- Identify at least five common features that a child can use to classify.
- Do object-picture and verbal/object-picture activities with children.

In math the term *set* is used to refer to any things which are put together to form one group. A set in math may contain from zero (an empty set) to an endless number of things. However, most sets which children work with have some observable limit. A set of dishes is usually for a certain number of place settings such as service for eight or service for twelve. A set of tires for a car is usually four plus a spare which equals five. A set of tires for a large truck consists of more than five tires.

To add is to put together or join sets. The four tires on the wheels plus the spare in the trunk equals five tires. To subtract is to separate a set into smaller sets. For example, one tire goes flat and is taken off. It is left for repairs, and the spare is put on. A set of one (the flat tire) has been taken away or subtracted from the set of five. There are now two sets: four good tires on the car and one flat tire being repaired.

Before doing any formal addition and subtraction, the child needs to learn about sets and how they can be joined and separated. That is, children must practice sorting (separating) and grouping (joining). This type of

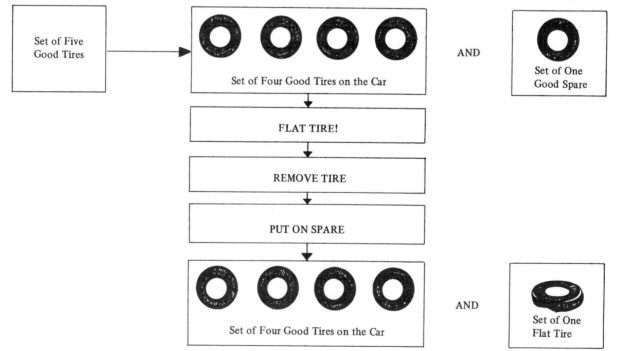

Fig. 7-1 Sets can be joined and separated.

activity is called *classification*. The child does tasks where he separates and groups things because they belong together for some reason. Things may belong together because they are the same color, or the same shape, do the same work, are the same size, are always together, and so on. For example, a child may have a box of wooden blocks and a box of toy cars (wood, metal, and plastic). The child has two sets: blocks and cars. He then takes the blocks out of the box and separates them by grouping them into four piles: blue blocks, red blocks, yellow blocks, and green blocks. He now has four sets of blocks. He builds a garage with the red blocks. He puts some cars in the garage. He now has a new set

of toys. If he has put only red cars in the garage, he now has a set of red toys. The blocks and toys could be grouped in many other ways by using shape and material (wood, plastic, and metal) as the basis for the groups. Figure 7-2 illustrates some possible groups using the blocks and cars.

Young children spend much of their playtime in just such classification activities. As the child works busily at these sorting tasks, he learns at the same time words which label what he has done. This happens when another person tells him the names and makes comments: "You have made a big pile of red things." "You have a pile of blue blocks, a pile of green blocks, ..." "Those are plastic

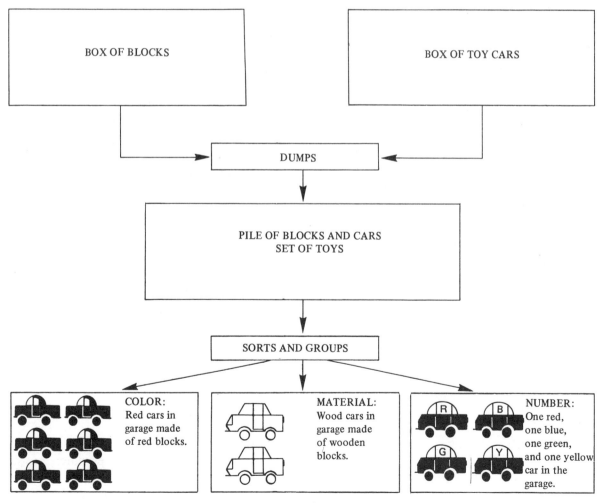

Fig. 7-2 Classification (Forming Sets) Through Play

and those are wood." As the child learns to speak, the adult questions him, "What color are these? Which ones are plastic?"

The child learns that things may be grouped together using a number of kinds of common features:

- **Color:** Things can go together that are the same color.
- **Shape:** Things may all be round, square, triangular, and so on.
- **Size:** Some things are big and some are small; some are fat and some are thin; some are short and some are tall.
- **Material:** Things are made out of different materials such as wood, plastic, glass, paper, cloth, and metal.
- **Pattern:** Things have different visual patterns such as stripes, dots, flowers, or may be plain (no design).
- **Texture:** Things feel different from each other (smooth, rough, soft, hard, wet, dry).
- **Function:** Some items do the same thing or are used for the same thing (all are for eating, writing, playing music, for example).
- **Association:** Some things do a job together (candle and match, milk and glass, shoe and foot) or come from the same place (bought at the store or seen at the zoo) or belong to a special person (the hose, truck, and hat belong to the fire fighter).
- **Class name:** There are names which may belong to several things (people, animals, food, vehicles, weapons, clothing, homes).
- **Common features:** All have handles or windows or doors or legs or wheels, for example.
- **Number:** All are groups of specific amounts (see unit 6) such as pairs, groups of three, four, five, and so on.

ASSESSMENT

The adult should note the child's play activities. Does he sort and group his play materials? For example, he might play with a pegboard and put each color peg in its own row; build two garages with big cars in one and small cars in another; when offered several kinds of crackers for snack, he might pick out only triangle shapes; he might say, "Only boys can be daddies — girls are mothers."

More formal assessment can be done using the tasks in the Appendix. Two examples are shown:

Sample Assessment Task   **Preoperational Ages 4-6**

Sets and Classifying (unit 7): Free Sort

Present the child with 20-25 assorted objects or pictures of objects and/or cutouts of shapes that can be grouped together by color, size, shape, or category (such as animals, people, furniture, clothing, or toys). PUT THE THINGS TOGETHER THAT BELONG TOGETHER. If the child looks puzzled, pick up an object and ask, "WHAT BELONGS WITH THIS?"

Sample Assessment Task   **Preoperational Ages 4-6**

Sets and Classifying (unit 7): Verbal Clue

Present the child with 20-25 objects or pictures of objects and/or cutouts of shapes that can be grouped together by color, shape, size, or category (such as animals, people, furniture, clothing, or toys). FIND SOMETHING THAT IS _____ . Name a specific color, shape, size, material, pattern, function, or class. Also ask, FIND SOME THINGS YOU CAN USE TOGETHER.

NATURALISTIC ACTIVITIES

Sorting and grouping is one of the most basic and natural activities for the young child. Much of his play is organizing and reorganizing the things in his world. The infant learns the set of people who take care of him most of the time (day care provider, mother, father, and/or relatives and friends) and others are put in the set of "strangers." He learns that some objects when pressed on his gums

Fig. 7-3 "Put the fruit in one bowl and the vegetables in another."

makes the pain of growing teeth less. These are his set of teething things.

As soon as the child is able to sit up, he finds great fun in putting things in containers and dumping them out. He can never have too many boxes, plastic dishes, and coffee cans along with safe items such as plastic beads and teething toys which can be put in and dumped out. This activity is his first try at building sets.

By age three the child sorts and groups to help organize his play activities. He sorts out from his things those which he needs for what he wants to do. He may pick out wild animal toys for his zoo; people dolls for his family play; big blocks for his house; blue paper circles to paste on paper; girls for friends, and so on.

The adult provides the free time, the materials (junk is fine as long as it is safe), and the space. The child does the rest.

INFORMAL ACTIVITIES

In informal ways the adult can let the child know that sorting and grouping activities are of value by showing that she likes what he is doing. This can be done with a look, a smile, a nod, or a comment.

In informal ways the adult can also build the child's classification vocabulary. She can label what the child has and ask questions

about what the child has done: "You have used all blue confetti in your picture." "You've used all the square blocks." "You have the pigs in this barn and the cows in that barn." "You painted green and purple stripes today." "Can you put the wild animals here and the farm animals here?" "Separate the spoons from the forks." "See if any of the cleaning rags are dry." "Put the crayons in the can and the pencils in the box." As the child has more words to use, he is able to describe in thought and out loud how he will sort and group. Words give him shortcuts for labeling sets.

STRUCTURED ACTIVITIES

The sorting and grouping which is the basis of classifying sets of things lends itself to many kinds of activities with many kinds of materials. As with matching activities (unit 5), real objects are used first, then pictures of real objects, cutouts, and patterns and symbols. The matching skill is needed in order to sort and group. Classification takes the child into higher levels of matching which includes more than one-to-one correspondence (putting one thing with one other thing). Several things may be classified into any one set. The child must keep in mind the basis for his group as he sorts through all available things. Remembering that when given three pigs and three

53

houses in separate piles that he must find a pig for each house is easier than being given a mixed pile of pigs and houses which he must first sort into a pile of pigs and a pile of houses before he can find if there is a house for each pig.

Structured activities to be used to learn to make sets are of two basic kinds: object-picture and verbal/object-picture.

- **Object-Picture Activities:** the child does not have to speak; only a motor response (point or pick something up) is needed.

  a. The child is shown a real object or a picture of an object and asked to choose from a bunch of objects or pictures the ones that go with the example. The example, chosen on the basis of any of the common features listed earlier in the unit, might be: a piece of fruit, an animal, something with a rough texture, something with a flower pattern on it. The basic question is FIND THINGS THAT GO WITH THIS.

  b. A bunch of things or pictures are given to the child. He must find his own basis to sort and group. The task is to FIND THINGS THAT BELONG (GO) TOGETHER.

- **Verbal/Object-Picture Activities:** the child must understand a verbal clue and may have to give a verbal answer.

  a. A verbal model is given: color, size, pattern, function, class, for example. The basic task is to find other objects or pictures which fit the requirement given. The task is to FIND THE THINGS THAT ARE (blue, round, for writing, animals, rough to touch, and so on).

  b. Show a group of things or pictures of things and ask, WHY DO THESE BELONG TOGETHER?

  c. Show a series of related things and ask the children to describe each one and tell why or how they are the same. For example, the teacher has a bag of things used to keep clean (soap, washcloth, toothbrush, etc.). She has the children describe the articles one at a time and then asks, WHAT IS THE SAME ABOUT THESE?

The following activities help children learn the idea of sets:

— SETS AND CLASSIFICATION: COLOR—

Objective: To sort and group by color

Materials: Several different objects that are the same color and four objects each of a different color; for example, a red toy car, a red block, a red bead, a red ribbon, a red sock, and so on, and one yellow car, one green ribbon, one blue ball, and one orange piece of paper

Activities:

1. Hold up one red object, FIND THE THINGS THAT ARE THE SAME COLOR AS THIS. After all the red things have been found: THESE THINGS ARE ALL THE SAME COLOR. TELL ME THE NAME OF THE COLOR. If there is no correct answer: THE THINGS YOU PICKED OUT ARE ALL RED THINGS. Ask: WHAT COLOR ARE THE THINGS THAT YOU PICKED OUT?

2. Put all the things together again: FIND THE THINGS THAT ARE *NOT* RED.

Follow-up: Repeat this activity with different colors and different materials. During free play put out a container of brightly colored materials. Note if the children put them into groups by color. If they do, ask, "Why did you put those together?" Accept any answer they give but note whether they give a color answer.

SETS AND CLASSIFICATION:
──────ASSOCIATION──────

Objective: To form sets of things that go together by association

Materials: Picture card sets may be bought or made. Each set can have a theme such as one of the following:

1. Pictures of people in various jobs and pictures of things that go with their job:

Worker	Things that go with the worker's job
letter carrier	letter, mailbox, stamps, hat, mailbag, mail truck
airplane pilot	airplane, hat, wings
doctor	stethoscope, thermometer, Band-Aid
trash collector	trash can, trash truck
police officer	handcuffs, pistol, hat, badge, police car
fire fighter	hat, hose, truck, boots and coat, hydrant, house on fire
grocer	various kinds of foods, bags, shopping cart

Start with about three sets and keep adding more.

2. Things that go together for use:

Item	Goes with
glass tumbler	carton of milk, pitcher of juice, can of soda pop
cup and saucer	coffee pot, teapot, steaming teakettle
match	candle, cigarette, campfire
paper	pencil, crayon, pen
money	purse, wallet, bank
table	four chairs

Start with three sets and keep adding more.

3. Things that are related, such as animals and their babies

Fig. 7-4 "The rake goes with the leaves."

Activities:

1. One at a time show the pictures of people or things which are the main clue (the workers for example) and ask: WHO (WHAT) IS THIS? When they have all been named, show the "go with" pictures one at a time: WHO (OR WHAT) DOES THIS BELONG TO?

2. Give each child a clue picture. Hold each "go with" picture up in turn: WHO HAS THE PERSON (OR THING) THIS BELONGS WITH? WHAT DO YOU CALL THIS?

3. Give a deck of cards to one child: SORT THESE OUT. FIND ALL THE WORKERS AND PUT THE THINGS WITH THEM THAT THEY USE. Or, HERE IS A GLASS, A CUP AND SAUCER, AND SOME MONEY. LOOK THROUGH THESE PICTURES AND FIND THE ONES THAT GO WITH THEM.

Follow-up: Have sets of cards available for children to use during free playtime. Note whether they use them individually or make up group games to play. Keep introducing more sets.

SETS AND CLASSIFICATION:
──────SIMPLE SORTING──────

Objective: To practice the act of sorting

Materials: Small containers such as margarine dishes filled with small objects such as buttons of various sizes, colors, and shapes, or with dried beans, peas, corn; another container with smaller divisions in it (such as an egg carton)

Activities:

1. Have the sections of the larger container marked with a model such as each kind of button or dried bean. The children match each thing from their container with the model until everything is sorted and grouped into new sets in the egg carton (or other large container with small sections).

2. Use the same materials but do not mark the sections of the sorting container. See how the child will sort on his own.

Follow-up: Have these materials available during free playtime. Make up more sets using different kinds of things for sorting.

SETS AND CLASSIFICATION: ─────CLASS NAMES, DISCUSSION─────

Objective: To discuss sets of things which can be put in the same class and decide on the class name

Materials: A set of things which can be put in the same group on the basis of class name, such as

1. animals: several toy animals

2. vehicles: toy cars, trucks, motorcycles

3. clothing: a shoe, a shirt, a belt, and so on

4. things to write with: pen, pencil, marker, crayon, chalk

Activities: The same plan can be followed for any group of things.

1. Bring the things out one at a time until three have been discussed. Ask about each.

   a. WHAT CAN YOU TELL ME ABOUT THIS?

   b. Five specific questions:
   WHAT DO YOU CALL THIS (WHAT IS ITS NAME?)
   WHAT COLOR IS IT?
   WHAT DO YOU DO WITH IT? or (WHAT DOES IT DO?) or (WHO USES THIS?)
   WHAT IS IT MADE OUT OF?
   WHERE DO YOU GET ONE?

   c. Show the three things discussed: WHAT DO YOU CALL THINGS LIKE THIS? THESE ARE ALL (ANIMALS, VEHICLES, CLOTHING, THINGS TO WRITE WITH, AND SO ON.)

2. Put two or more groups of things together that have already been discussed. Have the children sort them into sets and tell the class name for each set.

Follow-up: Put together sets like the above that include things from science and social studies.

SETS AND CLASSIFICATION: ─────LEARNING THE NAME *SET*─────

Objective: To learn the meaning of the term *set*

Materials: After the children have had many sorting and grouping experiences, use materials that are familiar (that they have used in sorting and grouping activities).

Activities:

1. Show the children groups of things they have already used, such as crayons, cups, buttons, toy cars, blocks, and so on. Tell them THIS IS A SET OF (OBJECT NAME). Show three or four sets of different things. When you have introduced several groups of things with THIS IS A SET OF (NAME OF OBJECTS), then point to each and ask: WHAT IS THIS? Always answer with YES, THIS IS A SET OF (OBJECT NAME) whether they say set or not. Next point to each set and ask, WHAT IS THE NAME OF THIS SET?

2. As soon as the child can use the name set to refer to groups, present bunches of objects or pictures of objects as done before, ask him to find the sets. Use bunches of objects which can be sorted into at least three different sets.

Follow-up: Use the term *set* whenever you use classification activities.

── SETS AND CLASSIFYING: BOOKS ──

Objective: To learn and discuss characteristics of sets using books

Materials: Books with themes centering around a group of things with a unifying feature (Many of Golden Press' Golden Shape Books are excellent and inexpensive.)

Animals

1. The Dog Book, 1964.
2. Jungle Babies, 1969.
3. The Cat Book, 1964.
4. The Bunny Book, 1965.
5. The Nest Book, 1968.

Vehicles

1. The Car Book, 1968.
2. The Truck and Bus Book, 1966.
3. The Boat Book, 1965.

Christmas

1. The Christmas Tree Book, 1966.

People

1. People in Your Neighborhood, 1971.
2. People in My Family, 1971.

Everyday Things

1. The Shopping Book, 1975 (stores).
2. My House Book, 1966 (rooms in a house).
3. The Telephone Book, 1968 (things in a house).
4. The Sign Book, 1968.
5. The Hat Book, 1965.
6. The Snowman Book, 1965 (winter).
7. The Apple Book, 1964 (fruit).

Activities:

1. Younger children enjoy looking at the pictures and labeling the items.
2. Read the books to young listeners.
3. Discuss the categories in each book. Children can compare with their own experience.
4. Bring real items that are pictured in book.
5. Use as parts of broader units.
6. When the child has learned the term *set* the teacher can ask him to look through the books and tell if they contain sets.

Follow-up: Have the books available on the bookshelf for children to look at during playtime.

EVALUATION

As the children play, note whether each one sorts and groups as part of his play activities. There should be an increase as the child grows and as he has more experiences with sets and classification activities. He should use more feature names when he speaks during work and play. He should use

color, shape, size, material, pattern, texture, function, association words, and class names.

1. Tim has a handful of colored candies. "First I'll eat the orange ones." He carefully picks out the orange candies and eats them one at a time. "Now, the reds." He goes on in the same way until all the candies are gone.

2. Diana plays with some small wooden animals. "These farm animals go here in the barn. Richard, you build a cage for these wild animals."

3. Mr. Flores tells Bob to pick out from a box of toys some plastic ones to use in the water table.

4. Mary asks the cook if she can help sort the clean tableware and put it away.

5. George and Sam build with blocks. George tells Sam, "Put the big blocks here, the middle-sized ones here, and the small blocks here."

For more formal evaluation, the sample assessment tasks and the tasks in the Appendix may be used.

SUMMARY

Sets are made up of things which are put together in one group. The act of putting things into groups by sorting out things which have some feature that is the same is called classification.

Classifying is a part of a child's normal play. He builds skills he will need later to add and subtract. He also adds to his store of ideas and words as he learns more features to be used to sort and to group.

Experiences which deal with sets and classifying may be called object-picture and verbal/object-picture.

SUGGESTED ACTIVITIES

- Put together a collection of objects of different sizes, shapes, colors, classes, materials, and uses. Put them in a large box or plastic dishpan. Present them in turn to a small group (two to six children each) of one-year-olds, two-year-olds, three-year-olds, four-year-olds, and five-year-olds. Tell the children, HERE ARE SOME THINGS YOU CAN PLAY WITH. Write a description of what each group does. Count the number of different features used for labeling and grouping by each age level. Compare the groups with each other.

- Add some set and classifying activities to your Activities File.

- Create a classification game. Try it out with one or more children. Report the results in class. Explain the game and tell about any changes that could be made to improve it. Ask the class for their comments.

- Visit an early childhood center. Note all the materials available for classification experiences. Write down a description of each classification activity observed. Which were naturalistic, which were informal, and which were structured? Interview the teacher. Find out how she incorporates classification activities into her daily program.

REVIEW

A. Define the term *set*.

B. Describe how you would explain the term *set* to young children.

C. Define the term *classification*.

D. Match the features in Column II with the behavior description in Column I.

Column I	Column II
1. The teacher says, "Find all the smooth objects."	a. color
2. The mother says, "Pour the milk in the glass."	b. shape
3. Mother says, "Put on your clothes."	c. size
4. The child says, "All the trees have leaves."	d. material
5. The teacher says, "Find all the things you can use to draw."	e. pattern
6. The children make a train with only yellow chairs.	f. texture
7. Mother has Tanya sort out the paper and cloth napkins.	g. function
8. Tom makes a pile of square blocks.	h. association
9. Teacher says, "Find three balloons for me."	i. class name
10. The child says, "These cookies are bigger than the ones on the table."	j. common features
11. "Tim," says mother, "you are wearing a checked shirt and striped pants."	k. number

unit 8 comparing

OBJECTIVES

After studying this unit, the student should be able to

- List and define comparison terms
- Identify the math ideas learned before comparing
- Do informal measurement and number comparing activities with children
- Do structured measurement and number comparing activities with children

When comparing, the child finds a relationship between two things or sets of things on the basis of some specific characteristic or attribute. One type of attribute is an informal measurement such as size, length, height, weight, or speed. A second type of attribute is number comparisons. To compare number, the child looks at two sets and decides if they have the same number of things or if one has more things. Comparing is the basis of ordering (unit 13) and measurement (unit 14 and unit 15).

Some examples of measurement comparisons are listed:

- John is taller than Mary.
- This snake is long. That worm is short.
- Father bear is bigger than baby bear.

Examples of number comparisons are shown:

- Does everyone have two gloves?
- I have more cookies than you have.
- We each have two dolls — that's the same.

THE BASIC COMPARISONS

To make comparisons and understand them, the child learns the following basic comparisons:

- Informal Measurement

large	small
big	little
long	short
tall	short
fat	skinny
heavy	light
fast	slow
cold	hot
thick	thin
wide	narrow
near	far
later	sooner (earlier)
older	younger (newer)
higher	lower
loud	soft (sound)

- Number

more	less

In addition, the child learns that some of the things compared may be *not different*, or the same. The child should be able to match, count, and classify at the lower levels before going into structured number comparison.

ASSESSMENT

During the child's play, the teacher should note any of the child's activities which might show he is comparing. For example, when a

Fig. 8-1 The *big* girl builds a *big* pile, and the *little* girl builds a *small* pile. The *big* girl is *taller* than her pile.

Fig. 8-2 "I make the blocks taller than me."

bed is needed for a doll and two shoe boxes are available, does he look the boxes over carefully and place the doll in each box in turn in order to get the doll into a box that is the right size? If he has two trucks, one large and one small, does he build a bigger garage for the large truck? The adult should also note with children old enough to talk if they use the words given in the list of basic comparisons.

In individual interview tasks, the child is asked questions to see if he understands and uses the basic comparison words. The child is presented with some objects or pictures of things which differ or are the same in size or number and is asked to tell if they are the same or different. Tasks (see Appendix) are like these shown:

Sample Assessment Task   **Preoperational Ages 4-5**

Comparing (unit 8): Informal Measurement

Present the child with two objects or pictures of objects which vary in size (height, length, or width):

   FIND (POINT TO) THE BIG BLOCK.
   FIND (POINT TO) THE SMALL BLOCK.

Fig. 8-3 "I can make it even higher."

Sample Assessment Task   *Preoperational Ages 3-4*

Comparing (unit 8): Number

Place two dolls (or toy animals or cutout figures) in front of the child. **I'M GOING TO GIVE EACH DOLL SOME COOKIES.** (Put two cardboard cookies in front of one doll and six cardboard cookies in front of the other). **SHOW ME THE DOLL THAT HAS MORE COOKIES.** Go through the procedure again asking **SHOW ME THE DOLL THAT HAS LESS (FEWER) COOKIES.**

Before giving the number comparison tasks, the teacher should be sure the child has begun to match, count, and classify.

NATURALISTIC ACTIVITIES

The young child has many contacts with comparisons in his daily life. At home mother says, "Get up, it's *late*. Mary was up *early*. Eat *fast*. If you eat slowly, we will have to leave before you are finished. Have a *big* bowl for your cereal, that one is too *small*." At school the teacher says, "I'll pick up this *heavy* box, you pick up the *light* one." "Sit on the *small* chair, that one is too *big*." "Let's finish this story." "Remember, the father bear's porridge was too *hot* and the mother bear's porridge was too *cold*."

As the child uses materials, he notices that things are different. The infant finds that some things can be grabbed and held because they are *small* and *light* while others cannot be held because they are *big* and *heavy*. As he crawls about, he finds he cannot go behind the couch because the space is too *narrow*. He can go behind the chair because there is a *wide* space between the chair and the wall. The young child begins to build with blocks and finds that there are *more small* blocks in his set of blocks than there are *large* ones. He notices that there are people in his environment who are big and people who are small in relation to him. One of the questions most often asked is "Am I a big boy?" or "Am I a big girl?"

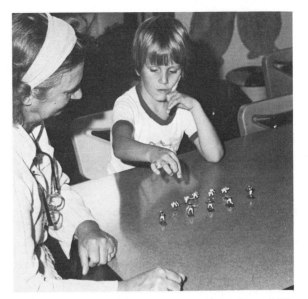

Fig. 8-4 "Are there more Browns or more Bengals?"

INFORMAL ACTIVITIES

Small children are very concerned about size and number, especially in relation to themselves. They want to be bigger, taller, faster, and older. They want to be sure they have the same, not less — and if possible more — of things than the other child has. These needs of the young child bring about many situations where the adult can help in an informal way to aid the child in learning the skills and ideas of comparing.

Informal measurements are made in a concrete way. That is the things to be compared are looked at, felt, lifted, listened to, and so on, and the attribute is labeled.

- Eighteen-month-old Brad tries to lift a large box of toy cars. Mr. Brown squats down next to him, holding out a smaller box of cars, "Here, Brad, that box is too big for your short arms. Take this small box."

- Three-year-olds, Kate and Chris, run up to Mrs. Raymond, "We can run fast. Watch us. We can run faster than you. Watch us." Off they go across the yard while Mrs. Raymond watches and smiles.

Fig. 8-5 "I'm high."

- Five-year-olds, Sam and George, stand back to back. "Check us, Mr. Flores. Who is taller?" Mr. Flores says, "Stand by the mirror and check yourselves." The boys stand by the mirror, back to back. "We are the same," they shout. "You are taller than both of us," they tell their teacher.

- It is after a fresh spring rain. The children are on the playground looking at worms. Comments are heard, "This worm is longer than that one." "This worm is fatter." Miss Collins comes up. "Show me your worms. Sounds like they are different sizes." "I think this small, skinny one is the baby worm," says Richard.

Comparative number is also made in a concrete way. When comparing sets of things just a look may be enough if the difference in number is large.

- "Teacher! Juanita has all the spoons and won't give me one!" cries Tanya.

If the difference is small the child will have to use his skill of matching (one-to-one correspondence). He may physically match each item, or he may count — depending on his level of development.

Fig. 8-6 "I'm low."

- "Teacher! Juanita has more baby dolls than I do." "Let's check," says Mr. Brown. "I already checked," says Tanya. "She has four and I have three." Mr. Brown notes that each girl has four dolls. "Better check again," says Mr. Brown. "Here, let's see. Put each one of your dolls next to one of Juanita's, Tanya." Tanya matches them up. "I was wrong. We have the same."

A child at a higher level of development could have been asked to count.

To promote informal learning, the teacher must put out materials that can be used by the child to learn comparisons on his own. The teacher must also be ready to step in and support the child's discovery by using comparison words and giving needed help with comparison problems which the child meets in his play and other activities.

STRUCTURED ACTIVITIES

Most children learn the idea of comparison through naturalistic and informal activities. For those who do not, more formal experiences can be planned. There are many commercial materials available individually and in kits which are designed to be used to teach comparison skills and words. Also, the environment is full of things that can be used.

The following are some basic types of activities which can be repeated with different materials.

COMPARISONS: INFORMAL ————————MEASUREMENTS ————

Objectives: To gain skill in observing differences in size, speed, temperature, age, and loudness. To learn the words associated with differences in size, speed, temperature, age, and loudness.

Materials: Use real objects first. Once the child can do the tasks with real things, then introduce pictures and chalkboard drawings.

Some Things Which Can Be Used	
Comparison	**Things to Use**
large-small and big-little	buttons, dolls, cups, plates, chairs, books, records, spools, toy animals, trees, boats, cars, houses, jars, boxes, people, pots, and pans
long-short	string, ribbon, pencils, ruler-meter stick, yardstick, snakes, worms, lines, paper strips
tall-short	people, ladders, brooms, blocks, trees, bookcases, flagpoles, buildings
fat-skinny	people, trees, crayons, animals, pencils, books, snowmen
heavy-light	same size but different weight containers (such as shoe boxes or coffee cans taped shut filled with items of different weights)
fast-slow	toy cars or other vehicles for demonstration, the children themselves — their own movements, cars on the street, music, talking
hot-cold	containers of water, food, ice cubes — boiling water, chocolate milk and hot chocolate, weather
thick-thin	paper-cardboard, books, pieces of wood, slices of food (bologna, cucumber, carrot), cookie dough
wide-narrow	streets, ribbons, paper strips, lines (chalk, crayon, paint), doorways, windows
near-far	children and fixed points in the room, places in the neighborhood, map
later-sooner (earlier)	arrival at school or home, two events
older-younger (newer)	People: babies, younger and older children, adults of different ages. Any things brought in that have not been in the environment before.
higher-lower	swings, slides, jungle gyms, birds in trees, airplanes flying, windows, stairs, elevators, balconies, shelves
loud-soft	voices: singing and talking, claps, piano, drums, records, doors slamming

Activities: The basic activity is to present the two things or events to be compared and ask the child to find or tell which one is which. Some examples are given:

- The teacher places two pieces of paper in front of the children. Each piece is one inch wide. One is six inches long and the other is twelve inches long. WHICH ONE IS LONGER (SHORTER)?

- The teacher places two identical coffee cans on the table. One is filled with sand; the other is empty. They are both taped closed so the children cannot see inside. PICK UP EACH CAN. TELL ME WHAT IS DIFFERENT ABOUT THEM.

The variety is almost endless in terms of experiences that can be offered with many kinds of things.

Follow-up: Set up a table with two empty containers and a third container holding pairs of opposite objects. One empty container should have a picture of one item and the other empty container a picture of an opposite item (such as a tall person and a short person). The child could then sort the pairs of things in the third container (such as clothing which fits the person) into the empty containers.

COMPARISONS: NUMBER

Objectives:
- To enable the child to compare sets of different sizes
- To enable the child to use the terms *more, less,* and *same number*

Materials: Any of the objects and things used for matching, counting, and classifying

Activities: The following basic activities can be done using many different kinds of materials.

1. Set up a flannel board with many felt shapes or a magnet board with many magnet shapes. Put up two groups of shapes: ARE THERE AS MANY CIRCLES AS SQUARES? (RED CIRCLES AS BLUE CIRCLES, BUNNIES AS CHICKENS)? WHICH SET HAS MORE? HOW MANY CIRCLES ARE THERE? HOW MANY SQUARES? The children can point, tell with words, and move the pieces around to show that they understand the idea.

2. Have cups, spoons, napkins, or food for snack or lunch: LET'S FIND OUT IF WE HAVE ENOUGH _____ FOR EVERYONE. Wait for the children to find out. If they have trouble, suggest they match or count.

3. Set up any kind of matching problems where one set has more things than the other: cars and garages, fire fighters and fire trucks, cups and saucers, fathers and sons, hats and heads, cats and kittens, animals and cages, and so on.

Fig. 8-7 "This is the short one. The other one is long."

Follow-up: Put out sets of materials which the children can use on their own. Go on to cards with pictures of different numbers of things which the children can sort and match. Watch for chances to present informal experiences:

- Are there more boys or girls here today?
- Do you have more thin crayons or more fat crayons?
- Do we have the same number of cupcakes as we have people?

EVALUATION

The teacher should note whether the child can use more comparing skills during his play and routine activities. Without disrupting his activity, the adult asks questions as the child plays and works:

- Do you have more cows or more chickens in your barn?
- (Child has made two clay snakes) Which snake is longer? Which is fatter?

- (Child is sorting blue chips and red chips into bowls) Do you have more blue chips or more red chips?
- (Child is talking about his family) Who is older, you or your brother? Who is taller?

The assessment tasks in the Appendix may be used for formal evaluation interviews.

SUMMARY

To compare is to find the relationship between two things or two groups of things. Comparing two things is making an informal measurement. Comparing two groups of things involves the use of matching, counting, and classifying skills to find out if sets have more, less, or the same number of things.

SUGGESTED ACTIVITIES

- Assess one or more children with the sample informal measurement and number tasks in this unit. On the basis of the results, plan activities at each child's level. Prepare the materials and carry out the activities. Note the children's responses. Compare your experiences with those of the other students in the class.
- Add comparison activities to your Activities File.
- Observe a group of four- and five-year-olds. Note each instance of the use of comparison words and comparison activities.
- Plan a comparison activity. Try it out on two classmates. Have them evaluate the presentation and suggest improvements.

REVIEW

A. Define and give examples of each of the following.

1. Comparing
2. Informal measurement attribute
3. Number measurement attribute

B. Match each item in Column II with the correct statement in Column I.

Column I	Column II
1. This stick is longer than yours.	a. Number comparison
2. My car goes faster.	b. Speed comparison
3. My baby is bigger than Mary's baby.	c. Weight comparison
4. You gave Pete one more push than you gave me.	d. Length comparison
5. This box is heavier than that one.	e. Height comparison
6. That man is taller than the lady.	f. Size comparison

C. Briefly answer each of the following.

1. What other math skills help the child make comparisons?
2. Explain two comparison assessment tasks.
3. List two examples of naturalistic comparison situations.
4. List two examples of informal comparison situations.
5. List two examples of structured comparison activities.

unit 9 shape

OBJECTIVES

After studying this unit, the student should be able to

- Describe naturalistic, informal, and structured shape activities for young children

- Assess and evaluate a child's knowledge of shape

- Help children learn shape through haptic, visual, and visual-motor experiences

Each thing in the environment has its own shape. Much of the play and activity of the infant during the sensorimotor stage centers on learning about shape. The infant learns through looking and through feeling with hands and mouth. Babies learn that some shapes are easier to hold than others. They learn that things of one type of shape will roll. They learn that some things have the same shape as others. Young children see and feel shape differences long before they can describe these differences in words. In the late sensorimotor and early preoperational stages, the child spends a lot of time matching and classifying things. Shape is often used as the basis for these activities.

Fig. 9-1 The children experiment with the shape matching toy.

As the child moves into the middle of the preoperational period, he can learn that there are some basic shapes (called geometric shapes) which have their own names. These are illustrated in figure 9-2. First the child learns to label circle, square, and triangle. Then he can learn rectangle, diamond, and oval. Later on, he will use these shape names in geometry. The purposes of learning specific shape names and finding shapes in pictures and in the environment are listed:

- The young child practices discriminating forms.

- The young child learns some names which he can use when describing things in the environment. ("I put the book on the square table.")

ASSESSMENT

Observational assessment can be done by noticing whether the child uses shape to organize his world. As the child plays with materials, the adult should note whether he groups things together because the shape is the same or similar. For example, a child plays with a set of plastic shape blocks. There are triangles, squares, and circles. Some are red, blue, green, yellow, and orange. Sometimes he groups them by color, sometimes by shape. A child

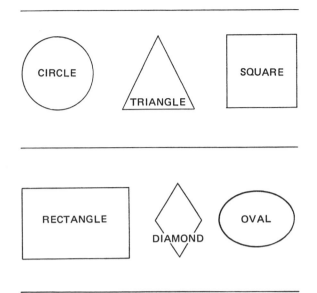

Fig. 9-2 Geometric Shapes

Sample Assessment Task   Preoperational Ages 5-6

Shape (unit 9): Environmental Geometry

LOOK AROUND THE ROOM. FIND AS MANY SHAPES AS YOU CAN....WHAT THINGS ARE SQUARE SHAPES?.....CIRCLES?.....RECTANGLES?....TRIANGLES?

NATURALISTIC ACTIVITIES

Naturalistic activities are most important in the learning of shape. The child perceives the idea of shape through sight and touch. The infant needs things to look at, to grasp, and to touch and taste. The toddler needs different things of many shapes to use as he sorts and matches. He needs many containers (bowls, boxes, coffee cans) and many objects (such as pop beads, ping pong balls, poker chips, and empty thread spools). He needs time to fill containers with these objects of different shapes and to dump the objects out and begin again. As he holds each thing, he examines it with his eyes, hands, and mouth.

The older preoperational child enjoys a junk box filled with things such as buttons, checkers, bottle caps, pegs, small boxes, and plastic bottles which he can explore. The teacher can also put out a box of attribute blocks (wood or plastic blocks in geometric shapes). Geometric shapes and other shapes

is playing with pop beads of different colors and shapes. Sometimes he makes strings of the same shape; sometimes, of the same color. The child may use some shape names in his everyday conversation.

The individual interview tasks for shape center on discrimination, labeling, matching, and sorting. Discrimination tasks assess whether the child can see that one form has a different shape from another form. Labeling tasks assess whether the child can find a shape when the name is given and whether he can name a shape when a picture is shown to him. At a higher level, he finds shapes in pictures and in his environment. Matching would require the child to find a shape like one shown to him. A sorting task would be one in which the child must separate a mixed bunch of shapes into sets. Tasks are like the samples shown:

Sample Assessment Task   Preoperational Ages 3-4

Shape (unit 9): Geometric Shape Recognition

With black marking pen draw a circle, a square, and a triangle each on a separate card. Place the cards in front of the child. **POINT TO THE SQUARE. POINT TO THE CIRCLE. POINT TO THE TRIANGLE.**

Fig. 9-3 The teacher finds out if the child knows the shape names.

Fig. 9-4 The child explores shape.

can also be cut from paper and/or cardboard and placed out for the child to use. Figure 9-5 shows some blob shapes that can be put into a box of shapes to sort.

In dramatic play, the child can put to use his ideas about shape. The preoperational child's play is representational. He uses things to represent something else which he does not have at the time. He finds something that is close to the real thing, and it is used to represent the real thing. Shape is usually one of the elements used when the child picks a representational object:

- A stick or a long piece of wood is used for a gun.

- A piece of rope or old garden hose is used to put out a pretend fire.

- The magnet board shapes are pretend candy.

- A square yellow block is a piece of cheese.

- A shoe box is a crib, a bed, or a house — as needed.

- Some rectangular pieces of green paper are dollars, and some round pieces of paper are coins.

INFORMAL ACTIVITIES

The teacher can let the child know that she notices his use of shape ideas in activities through comments and attention. She can also supply him with ideas and objects which will fit his needs. She can suggest or give the child a box to be used for a bed or a house, some blocks or other small objects for his pretend food, or green rectangles and gray and brown circles for play money.

Labels can be used during normal activities. The child's knowledge of shape can be used too.

- "The forks have sharp points; the spoons are round and smooth."

- "Put square placemats on the square tables and rectangular placemats on the rectangular tables."

- "We'll have triangle shaped crackers today."

Fig. 9-5 Blob Shapes: Make up your own.

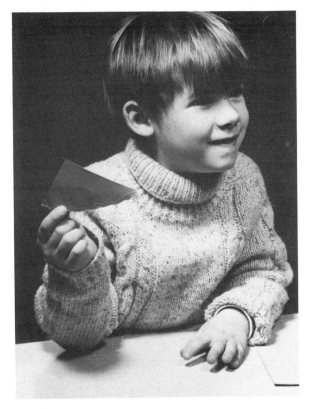

Fig. 9-6 "I tore a triangle from this paper."

- As a child works on a hard puzzle, the teacher takes his hand and has him feel the empty space with the index finger, "Feel this shape and look at it. Now find the puzzle piece that fits here."

- As the children use clay or play dough, the teacher says, "You are making lots of shapes. Kate has made a ball; Jim, a snake; and Diana, a pancake."

- During cleanup time, the teacher says, "Put the square blocks here and the rectangle blocks here."

The teacher can pay attention and respond when the child calls her attention to shapes in the environment. The following examples show that children can generalize; they can use what they know about shape in new situations.

- "Ms. Moore, the door is shaped like a rectangle." Ms. Moore smiles and looks over at George, "Yes, it sure is."

- "The plate and the hamburger look round like circles." "They do, don't they," comments Mr. Brown.

- "Where I put the purple paint, it looks like a butterfly." Mr. Flores looks over and nods.

- "The roof is shaped like a witch's hat." Miss Conn smiles.

- Watching a variety show on TV, the child asks, "What are those things that are shaped like bananas?" (Some curtains over the stage are yellow and do look just like big bananas!) Dad comments laughingly, "That is funny. Those curtains look like bananas."

STRUCTURED ACTIVITIES

Structured shape activities involve two main operations: *discrimination* (seeing or feeling that one shape is the same as or different from another) and *labeling* (giving a name to shapes which are seen and/or felt). Children need both haptic and visual experiences to

Fig. 9-7 The teacher observes as the child puts the shapes on the matrix board.

Fig. 9-8 "This is a circle."

learn discrimination and labeling. These experiences can be described as follows:

- *Haptic activities* use the sense of touch to match and identify shapes. These activities involve experiences where the child cannot see to solve a problem but must use only his sense of touch. The items to be touched are hidden from view. Older children can be blindfolded. The things may be put in a bag or a box or wrapped in cloth or paper. Sometimes a clue is given. The child can feel one thing and then find another that is the same shape. The child can be shown a shape and then asked to find one that is the same. Finally, the child can be given just a name (or label) as a clue.

- *Visual activities* use the sense of sight. The child may be given a visual or a verbal clue and asked to choose from several things the one that is the same shape. Real objects or pictures may be used.

- *Visual-motor activities* use the sense of sight and motor coordination at the same time. This type of experience includes the use of puzzles, formboards, attribute blocks, flannel boards, magnet boards, colorforms, and paper cutouts which the child moves about on his own. He may sort the things into sets or arrange them into a pattern or picture. Sorting was described in unit 7. Examples of making patterns or pictures are shown in figure 9-9.

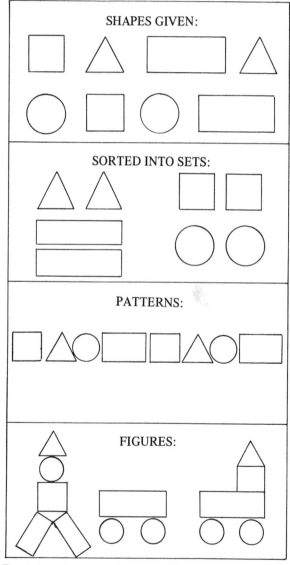

Fig. 9-9 Shapes can be sorted into sets, put into a pattern, or made into figures.

As the child does haptic, visual, and visual-motor activities, the teacher can use labels (words such as round, circle, square, triangle, rectangle, shape). The following activities are some examples of basic types of shape experiences for the young child.

--------------SHAPE: FEELING BOX--------------

Objective: To give the child experiences which will enable him to use his sense of touch to label and discriminate shapes

Materials: A medium-sized cardboard box covered with plain Contac® paper with a hole cut in the top big enough for the child to put his hand in but small enough so the child cannot see inside; some familiar objects, such as a toy car, a small wooden block, a spoon, a small coin purse, a baby shoe, a pencil, and a rock

Activities:

1. Show the child each of the objects. Be sure he knows the name of each one. Have him pick each one up and name it.

2. Out of his sight, put the objects in the box.

3. The following can then be done:

   - Have another set of identical objects. Hold them up one at a time: PUT YOUR HAND IN THE BOX. FIND ONE LIKE THIS.

   - Have a set of identical objects. Put each one in an individual bag: FEEL WHAT IS IN HERE. FIND ONE JUST LIKE IT IN THE BIG BOX.

   - Use just a verbal clue: PUT YOUR HAND IN THE BOX. FIND THE ROCK (CAR, BLOCK, ETC.)

   - PUT YOUR HAND IN THE BOX. TELL ME THE NAME OF WHAT YOU FEEL. BRING IT OUT AND WE'LL SEE IF YOU GUESSED IT.

Follow-up: Once the child understands the idea of the "feeling box," a "mystery box" can be introduced. In this case, familiar objects are placed in the box, but the child does not know what they are. He must then feel them and guess what they are. Children can take turns. Before the child takes the object out, encourage him to describe it (smooth, rough, round, straight, bumpy, it has wheels, and so on). After the child learns about geometric shapes, the box can be filled with cardboard cutouts or attribute blocks.

SHAPE: DISCRIMINATION OF --------------GEOMETRIC SHAPES --------------

Objective: To see that geometric shapes may be the same or different from each other.

Materials: Any or all of the following may be used:

- Magnet board with magnet shapes of various types, sizes, and colors

- Flannel board with felt shapes of various types, shapes, and colors

- Attribute blocks (blocks of various shapes, sizes, and colors)

- Cards with pictures of various geometric shapes in several sizes (They can be all outlines or solids of the same or different colors.)

Activities: The activities are matching, classifying, and labeling.

- Matching: Put out several different shapes. Show the child one shape, FIND ALL THE SHAPES LIKE THIS ONE.

- Classifying: Put out several different kinds of shapes. PUT ALL THE SHAPES THAT ARE THE SAME KIND TOGETHER.

- Labeling: Put out some shapes — several kinds. Then ask, FIND ALL THE TRIANGLES (SQUARES, CIRCLES, ETC.) or TELL ME THE NAME OF THIS SHAPE. (Point to one at random.)

Follow-up: Do individual and small group activities. Do the same basic activities with different materials.

SHAPE: DISCRIMINATION AND --------------MATCHING GAME --------------

Objective: To practice matching and discrimination skills (for the child who has had experience with the various shapes already)

Materials: Cut out some shapes from cardboard. The game can be made harder by the number of shapes used, the size of the shapes, and the number of colors. Make six Bingo-type cards (each one should be different) and a spinner card which includes all the shapes used:

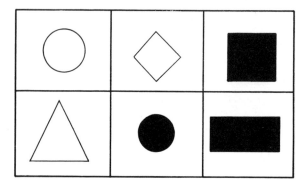

Cut out some plain squares of paper or use chips or buttons for markers.

Activity:

1. Give each child a Bingo card.
2. Have each child in turn spin the spinner. If he has the shape on his card which the spinner points to, he can cover the shape with a paper square or put a marker on it.

Follow-up: Once the rules of the game are learned the children can play it on their own.

–SHAPE: ENVIRONMENTAL GEOMETRY–

Objective: To see that there are geometric shapes all around in the environment

Materials: The classroom, the school building, the playground, the home, and the neighborhood

Activities:

1. Look for shapes on the floor, the ceiling, doors, and windows, materials, clothing, trees, flowers, vehicles, walls, fences, sidewalks, and so on.
2. Make a shape table. Cover the top and divide it into sections. Mark each section with a sample shape.

   Have the children bring things from home and put them on the place on the table that matches the shape of the thing that they bring.

SHAPE: FINDING SHAPES IN PICTURES AND DESIGNS

Objective: To sharpen discrimination skills by finding shapes in pictures

Materials: Picture books (See list in unit 7), pictures from workbooks or drawn by the teacher, figure 9-10.

Activity: Use the books and pictures with a child or a small group of children. Say, TELL ME WHAT YOU SEE IN THE PICTURE. Note whether any geometric shapes or other shapes which are in the picture are mentioned. Say, FIND THE _____ IN THE PICTURE. HOW MANY ARE THERE?

Follow-up: Put a large "find the shape" picture on the bulletin board so children can work on it on their own.

EVALUATION

Through observing during free play and during structured experiences, the teacher can see whether the child shows an increase in ideas regarding shape. She observes whether the child uses the word shape and other shape words as he goes about his daily activities. When he sorts and groups materials, the teacher notices whether he sometimes uses shape as the basis for organizing. The adult

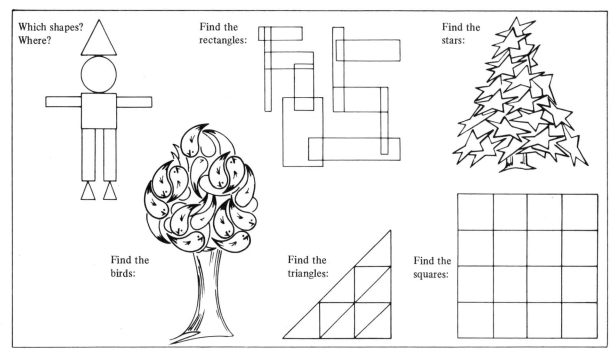

Fig. 9-10 Find the shapes.

gives the child informal tasks such as "Put the box on the square table"; "Fold the napkins so they are rectangle shapes"; Find two boxes that are the same shape."

After a period of instruction, the teacher may use formal interview tasks such as those in the Appendix.

SUMMARY

Each thing the child meets in the environment has shape. The child explores his world and learns in a naturalistic way about the shape of each object in it. Adults help by giving the child things to view, hold, and feel. Adults also teach the child words which describe shapes and the names of geometric shapes such as square, circle, and triangle.

The process of seeing that some shapes are the same and some are different is like the one the child uses later to see that some number and letter symbols are the same and some are different.

SUGGESTED ACTIVITIES

- Assess a child's shape ideas. Use the sample assessment tasks. Plan an activity at the child's level. Prepare the materials, and use the activity with the child. Evaluate the results.

- Put together a set of materials for a haptic experience. Try them out on the rest of the class. Have the class members rank the materials as to their value for this sort of activity. Put a list of the best on a 5″ x 8″ card for the Activities File.

- Maria Montessori stressed haptic activities. Find one of her books (or a book about her method) in the library. Visit a Montessori classroom. Note how well Montessori's ideas are used. Try some of the materials. Write an evaluation of this aspect of Montessori.

- Obtain permission from the director to take some haptic, visual, and visual-motor activities to a preschool center. Set those out for the children to explore. If some children seem interested, direct them in the more structured activities. Share with the class what was learned from this experience.

REVIEW

A. Below is a description of four-year-old Tim's activities on a school day. List the thirteen shape activities which are described and indicate whether the activity is naturalistic, informal, or structured.

Tim's mom calls him at 7:30 A.M. and says "It's time to get ready for school, Tim." Tim stretches and squeezes his pillow. He looks at his pillow and snuggles his head into its softness. Mom calls again. She comes into his room. "Come on, Tim. Get up and go wash your face." They walk into the bathroom. Mother gets out a clean washcloth and towel. "Now wash good," she says. "I will get breakfast. What do you want, eggs or cereal?" Tim says, "I want one of those round waffles." Mom says "O.K."

Tim takes the washcloth and splashes in the water. He tries to make bubbles from the bar of soap. He pushes the soap to the bottom of the sink. He watches it "pop" back up. He pushes it again. He becomes frustrated at the way it slides through his fingers. He grabs it with both hands and throws it back into the water. Splash! He delights in the splashing. Mother comes back. "Oh, Tim, what a mess. Come on, I told you to wash. It's not time to play with the soap and water. We have to hurry this morning. I have an early meeting." They go into the kitchen. Mom says, "Here is your round waffle. Look, I put it on a square plate. Here's some milk. Now eat, then we'll get dressed."

After Mom helps Tim get dressed, they get in the car and head toward the day care center. Tim looks out the window. He sees apartment buildings, signs, other cars, and buses.

At the day care center, Tim is welcomed by the director. "Good morning Tim." "Good morning Mrs. Adams. How are you today? It sure is getting cold."

Tim's day begins. He looks around the room and runs over by some other boys. They are playing with the unit blocks. Tim starts to build also. He uses long blocks, rectangular blocks, square blocks, and curved blocks. He makes a square enclosure. Miss Collins comes over. She says, "What are you going to put in your square building, Tim?" Tim says, "My truck."

Later Tim goes over to the art activity center. He takes some pipe cleaners and bends and twists them. He makes circles and glues them on some paper. "Look, Teacher, I made circle people."

For snack the children have cubes of cheese and round crackers. Then Tim plays with the wooden puzzles. The puzzles contain many different shapes. Miss Collins says, "Children, come on, it's story time. Today we're going to learn about magic squares, circles, triangles, and rectangles." As Miss Collins reads the story, she stops to ask questions about the shapes. After the story she puts on a record, and the children play a musical shape game. After this game, the children are told to go play with some of the new materials placed in the discovery center. Tim takes out the attribute blocks. He stacks all the same sized circles in a pile. Next, he stacks the squares. "Look John, I made a pile of circles and a pile of squares."

B. Answer each of the following questions.

1. What is the best way to assess a child's use of shape?

2. Describe one shape assessment task.

3. Describe one structured shape activity.

4. Define each: discrimination, labeling, matching, and sorting.

C. Match the shape tasks in Column II with the correct shape activity in Column I.

Column I

1. Teacher shows child a circle and says, "What is this shape called?"

2. Teacher says, "Point to the circle."

3. Teacher has a feely box of shapes.

4. Child is shown a circle and told to find one other object just like it.

5. Child plays with shape form fitting box.

6. Children play shape Bingo.

Column II

a. Shape discrimination task

b. Shape labeling task

c. Shape matching task

d. Shape sorting task

unit 10 space

OBJECTIVES

After studying this unit, the student should be able to

- Define the five space ideas and tell how each answers specific questions
- Assess and evaluate a child's ideas about space
- Do informal and structured space activities with young children

Space is a part of geometry just as is shape (unit 9). There are relationships in space and there is the use of space. The relationship ideas are position, direction, and distance. Use of space includes organization and patterns and construction. Each space idea helps the child answer his own questions:

Space Idea	Question	Answers
Position	Where (am I, are you, is he)?	on-off; on top of-over-under; in-out; into-out of; top-bottom; above-below; in front of-in back of-behind; beside-by-next to; between
Direction	Which way?	up-down; forward-backward; around-through; to-from; toward-away from; sideways; across
Distance	What is the relative distance?	near-far; close to-far from
Organization and pattern	How can things be arranged so they fit in a space?	arrange things in the space until they fit or until they please the eye
Construction	How is space made? How do things fit into the space?	Arrange things in the space until they fit; change the size and shape of the space to fit what is needed for the things

ASSESSMENT

A great deal about the child's ideas of space can be learned through observation. The adult notes the child's use of space words: Does he respond with an appropriate act when he is told the following?

- Put the book **on** the table.
- Please, **take off** your hat.
- You'll find the soap **under** the sink.
- Stand **behind** the gate.
- Sit **between** Kate and Chris.
- Move **away from** the hot stove.
- It's on the table **near** the window.

Fig. 10-1 "Room for me too!" — The toddler learns about space.

Fig. 10-2 John is between two friends.

Does he answer space questions using space words?

• Where is the cat? **On** the bed.

• Where is the cake? **In** the oven.

• Which way did John go? He went **up** the ladder.

• Where is your house? **Near** the corner.

The adult should note the child's use of organization and pattern arrangement during his play activities:

• When he does artwork, such as a collage, does he take time to place the materials on the paper in a careful way? Does he seem to have a design in mind?

• Does the child's drawing and painting show balance? Does he seem to get everything into the space that he wants to have it in, or does he run out of space?

• As he plays with objects, does he place them in straight rows, circle shapes, square shapes, and so on?

The teacher should note the child's use of construction materials such as blocks and containers:

• Does the child make structures with small blocks that toys such as cars and animals can be put into?

• Does he use the large blocks to make buildings that large toys and children can get into?

• Can he usually find the right size container to hold things (such as a shoe box that makes the right size bed for his toy bear)?

The teacher should note the child's use of his own body in space:

• When he needs a cozy place in which to play, does he choose one that fits his size, or does he often get stuck in tight spots?

• Does he manage to move his body without too many bumps and falls?

The individual interview tasks for space center on relationships and use of space. The tasks presented in the formal interview are like the following examples:

Sample Assessment Task   Preoperational Ages 2-3
Space (unit 10): Position
Have a small container (box, cup, or bowl) and an object such as a coin, a checker, or a chip. **PUT THE** *object name* **IN THE BOX (or cup or dish). Repeat using other space words (ON, OFF, OUT, IN FRONT OF, NEXT TO, UNDER) in place of the space word IN.**

Sample Assessment Task   Preoperational Ages 3-4

Space (unit 10): Position

Have some small containers and several small objects — for example, four plastic glasses and four small toy animals such as a fish, a dog, a cat, and a mouse. Ask the child to name each animal (or other object) so that you can use his name for each object if it is different from yours. Line up the glasses. Place the animals so one is *in*, one *on,* one *under*, and one *between* the glasses. Then say: TELL ME, WHERE IS THE FISH? WHERE IS THE DOG? WHERE IS THE CAT? WHERE IS THE MOUSE? Note whether the child uses the position words in his answer.

NATURALISTIC ACTIVITIES

It is through his everyday motor activities that the child first learns about space. As he moves his body in space, he learns position, direction, and distance relationships and about the use of the space. Children in the sensorimotor and preoperational stages need equipment which lets them place their own bodies on, off, under, over, in, out, through, above, below, and so on. They need places to go up and down, around and through, and sideways and across. They need things which they can put in, on, and under other things. They need things that they can place near and far from

Fig. 10-3 "Put the people on the cups."

other things. They need containers of many sizes to fill, blocks with which to build, and paint, collage, wood, clay and such which can be made into patterns and organized in space. Thus, when the child is matching, classifying, and comparing, he is learning about space at the same time.

The child who crawls and creeps often goes under furniture. At first he sometimes gets stuck when he has not judged correctly the size space under which he will fit. As he begins to pull himself up, he tries to climb on things. This activity is important not only for his motor development but for his spatial learning. However, many pieces of furniture are not safe or are too high. An empty beverage bottle box with the dividers still in it may be taped closed and covered with some colorful Contac® paper. This makes a safe and inexpensive place to climb. The adults can make several, and the child will have a set of large construction blocks. It has been stated in other units that the child needs safe objects to aid in developing math ideas and skills. Each time the child handles an object, he may learn more than one skill or idea. For instance, Juanita builds a house with some small blocks. The blocks are different colors and shapes. First Juanita picks out all the square, blue blocks and piles them three high in a row. Next she picks all the red rectangles and piles them in another direction in a row. Next she piles orange rectangles to make a third side to her structure. Finally, she lines up some yellow cylinders to make a fourth side. She places two pigs, a cow, and a horse in the enclosure. Juanita has sorted by color and shape. She has made a structure with space for her farm animals (a class) and has put the animals *in* the enclosure.

INFORMAL ACTIVITIES

Space is an area where there are many words to be learned and attached to actions.

Fig. 10-4 **"Where can I find a parking space for this truck?"**

The teacher should use space words (as listed earlier in the unit) as they fit into the daily activities. She should give space directions, ask space questions, and make space comments. Examples of directions and questions are in the assessment section of this unit. Space comments would be such as the following:

- Bob is at the *top* of the ladder.
- Cindy is *close* to the door.
- You have the dog *behind* the mother in the car.
- Tanya can move the handle *backward* and *forward*.
- You children made a house big enough for all of you. (construction)
- You pasted all the square shapes on your paper. (organization and pattern)

Jungle gyms, packing crates, ladders, ramps and other equipment designed for large muscle activity give the child experiences with space. They climb *up, down*, and *across*. They climb *up* a ladder and crawl or slide *down* a ramp. On the jungle gym they go *in* and *out*, *up* and *down*, *through*, *above* and *below* and *around*. They get *in* and *out* of packing crates. On swings they go *up* and *down* and *backward* and *forward* and see the world *down below*.

With the large blocks, boxes, and boards they make structures that they can get *in*

themselves. Chairs and tables may be added to make a house, train, airplane, bus, or ship. Props such as a steering wheel, fire fighter or police hats, ropes, hoses, discarded radios, and so on inspire children to build structures on which to play. With small blocks, children make houses, airports, farms, grocery stores, castles, trains, and so on. They then use their toy animals, people, and other objects in spatial arrangements, patterns, and positions. They learn to fit their structures into space available: on the floor, on a large table, or on a small table.

As the child works with art materials, he plans what to choose to glue or paste on his paper. A large selection of collage materials such as scrap paper, cloth, feathers, plastic bits, yarn, wire, gummed paper, cotton balls, bottle caps, and ribbon offer a child a choice of things to organize on the space he has. As he gets past the first stages of experimentation, the child plans his painting. He may paint blobs, geometric shapes, stripes, or realistic figures. He enjoys printing with sponges or potatoes. All these experiences add to his ideas about space.

Fig. 10-5 **Children learn about the position of their bodies in space.**

Fig. 10-6 Building a Structure from Paste and Scrap Materials

STRUCTURED ACTIVITIES

Structured activities of many kinds can be done to help the child with his ideas about space and his skills in the use of space. Basic activities are described for the three kinds of space relations (position, direction, and distance) and the two ways of using space (organization/pattern and construction).

SPACE: RELATIONSHIPS, PHYSICAL SELF

Objective: To relate the child's self to his position in space relative to other people and things

Materials: The child's own body, other people, and things in the environment

Activities:

1. Obstacle Course
   Set up an obstacle course using boxes, boards, ladders, tables, chairs, and like items. Set it up so that by following the course, the children can physically experience position, direction, and distance. This can be done indoors or out-

doors. As the child proceeds along the course, use space words to label his movement: "Leroy is going *up* the ladder, *through* the tunnel, *across* the bridge, *down* the slide and *under* the table. Now he is *close to* the end."

2. Find Your Friend
   Place children around in different places: sitting or standing on chairs or blocks or boxes, under tables, sitting three in a row on chairs facing different directions, and so on. Have each child take a turn to find a friend:
   FIND A FRIEND WHO IS ON A CHAIR (A BOX, A LADDER).
   FIND A FRIEND WHO IS UNDER A TABLE (ON A TABLE, NEXT TO A TABLE).
   FIND A FRIEND WHO IS BETWEEN TWO FRIENDS (BEHIND A FRIEND, NEXT TO A FRIEND).
   FIND A FRIEND WHO IS SITTING BACK-WARDS (FORWARDS, SIDEWAYS).
   FIND A FRIEND WHO IS ABOVE ANOTHER FRIEND (BELOW ANOTHER FRIEND).
   Have the children think of different places they can place themselves. When they know the game let the children take turns saying the FIND statements.

3. Put Yourself Where I Say
   One at a time give the children instructions for placing themselves in a position:
   CLIMB UP THE LADDER.
   WALK BETWEEN THE CHAIRS.
   STAND BEHIND TANYA.
   GET ON TOP OF THE BOX.
   GO CLOSE TO THE DOOR (GO FAR FROM THE DOOR)
   As the children learn the game, they can give the instructions.

4. Where Is Your Friend?
   As in activity number two, *Find Your Friend*, place the children in different places. This time ask WHERE questions. The child must answer in words. Ask WHERE IS (Child's Name)? Child answers, "Tim is under the table," or "Mary is on top of the playhouse."

Follow-up: Set up obstacle courses for the children to use during free playtime both indoors and outdoors.

—SPACE: RELATIONSHIPS, OBJECTS —

Objective: To be able to relate the position of objects in space to other objects

Materials: Have several identical containers (cups, glasses, boxes) and some small objects such as blocks, pegs, buttons, sticks, toy animals, people

Activities:

1. Point To
   Place objects in various spatial relationships such as shown below:

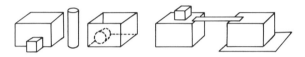

   POINT TO (OR SHOW ME) THE THING THAT IS (IN, ON, UNDER, BETWEEN, BEHIND, etc.) A BOX.

2. Put The
   Set some containers out. Place some objects to the side. Tell the child PUT THE <u>object name</u> (IN, ON, THROUGH, ACROSS, UNDER, NEAR) THE CONTAINER.

3. Where Is?
   Place objects as in "1" above and/or around the room. Ask, WHERE IS object? TELL ME WHERE THE <u>object</u> IS. Child should reply using space word.

Follow-up: Repeat the activity using different objects and containers. Leave the materials out for the children to use during free play.

SPACE: USE, ORGANIZATION/
—————PATTERN—————

Objective: To organize materials in space in a pattern

Materials: Many kinds of materials are available which will give the child experiences in making patterns in space. Some of these are listed:

   1. Geoboards are square boards with attached pegs. Rubber bands of different colors can be stretched between the pegs to form patterns and shapes.

Fig. 10-7 The geoboard is used to organize patterns in space.

2. Parquetry and pattern blocks are blocks of various shapes and colors which can be organized into patterns.

3. Pegboards are boards with holes evenly spaced. Individual pegs can be placed in the holes to form patterns.

4. Color inch cubes are cubes with one-inch sides. They come in sets with red, yellow, blue, green, orange, and purple cubes.

Fig. 10-8 Parquetry blocks can also be organized into patterns.

Activities:

1. The children can experiment freely with the materials and create their own patterns.

2. Patterns can be purchased or made for the children to copy.

Follow-up: After the children have been shown how they can be used, the materials can be left out for use during free playtime.

SPACE: USE, ORGANIZATION/ ────── PATTERN ──────

Objective: To organize materials in space in a pattern

Materials: Construction paper, scissors, and glue

Activity: Give the child one 8 1/2" x 11" piece of construction paper. Give him an assortment of precut construction paper shapes (squares, rectangles, triangles). Show him how many kinds of patterns can be made and glued on the big piece of paper. Then tell him to make his own pattern.

Follow-up: Offer the activity several times. Use different colors for the shapes, use different sizes, and change the choice of shapes.

──── SPACE: USE, CONSTRUCTION ────

Objective: To organize materials in space in three dimensions through construction

Materials: wood chips, polythene, cardboard, wire, bottle caps, and other scrap materials, Elmer's glue, heavy cardboard, or scraps of plywood

Activity: Give the child a bottle of glue and a piece of cardboard or plywood for a base. Let him choose from the scrap materials things to use to build a structure on the base. Encourage him to take his time, plan, and choose carefully which things to use and where to put them.

Follow-up: Keep plenty of scrap materials on hand so that children can make structures when they are in the mood.

──── SPACE: USE, CONSTRUCTION ────

Objective: To organize materials in space in three dimensions through construction

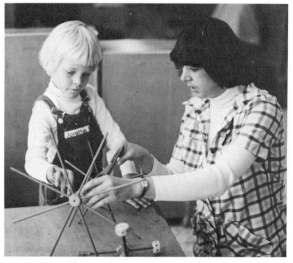

Fig. 10-9 Tinker Toys are made into a three-dimensional structure.

Materials: Many kinds of construction materials can be purchased which help the child to understand space and also improve eye-hand coordination and small muscle skills. Some of these are listed:

1. Lego: Jumbo for the younger child, regular for the older child or one with good motor skills

2. Tinker Toys

3. Bolt-it

4. Snap-N-Play blocks

5. Rig-A-Jig

6. Octons, Play Squares (and other things with parts that fit together)

Activities: Once the child understands the ways that the toys can be used, he can be left alone with the materials and his imagination.

EVALUATION

Informal evaluation can be done through observation. The teacher should note the following as the children proceed through the day:

- Does the child respond to space words in a way that shows understanding?

- Does he answer space questions and use the correct space words?

- Does his artwork and block building show an increase in pattern and organization?
- Does the child handle his body well in space?
- Does his use of geoboards, parquetry blocks, color inch cubes, and/or pegboards show an increase in organization and pattern?

After a period of instruction, the teacher should use the formal interview tasks in the Appendix.

SUMMARY

Space is an important part of geometry. The child needs to understand the relationship between his body and other things. He must also understand the relationship between things around him. Things are related through position, direction, and distance.

The child also needs to be able to use space in a logical way. He learns to fit things into the space available and to make constructions in space.

SUGGESTED ACTIVITIES

- Observe some children as they play in an early childhood center. Note instances of behaviors which reflect a child's feelings, ideas, and use of space as listed in the assessment section. Rank the children from those with well developed ideas about space to those whose ideas are not so well developed.

- Design a jungle gym/climbing structure that would offer all types of spatial experiences. Explain what kind of spatial activity can be experienced on each part.

- Look through some early childhood materials catalogues. Pick out and list the pattern and construction toys that could be purchased with $100 to spend.

- Add five space activities to your Activities File.

- Use four space assessment tasks to interview a three-year-old, a four-year-old and a five-year-old. Compare the response of the three children. Discuss the results with other students who try out the assessment tasks. Did they get similar results?

REVIEW

A. Answer each of the following questions.

1. How can a teacher assess a child's ideas about space through observation?

2. What are three informal space activities?

3. What are three structured space activities?

B. Select an item in Column II which applies to each of the phrases in Column I.

<table>
<tr><td align="center">Column I</td><td align="center">Column II</td></tr>
</table>

1. Answers the question, "What's the relative distance?"
2. Lincoln logs
3. How can things be arranged as they fit in space?
4. When a child arranges things in space until they fit
5. Space relations
6. Near-far
7. Parquetry blocks
8. "Here We Go 'Round the Mulberry Bush."
9. Answers the question, "Where am I?"
10. Answers the question, "Which way?"
11. Geoboards
12. Use of space
13. Over-under
14. Answers the question, "How is space made?"
15. Up-down

a. Position
b. Direction
c. Distance
d. Organization
e. Construction

unit 11 parts and wholes

OBJECTIVES

After studying this unit, the student should be able to

- Describe the three types of part/whole relationships
- Assess and evaluate a child's knowledge of parts and wholes
- Do informal and structured part/whole activities with young children

The young child learns that wholes have parts. Later the older child learns that parts are *fractions* of the whole. The child must learn the idea of parts and wholes before he can understand fractions. He learns that some things are made up of special (unique) parts, that sets of things can be divided into parts, and that whole things can be divided into smaller parts.

He learns about special parts:

- Bodies have parts (arms, legs, head).
- A car has parts (engine, doors, steering wheel, seats).
- A house has parts (kitchen, bathroom, bedroom, livingroom).
- A chair has parts (seat, legs, back).

He learns that sets of things can be divided:

- He passes out cookies for snack.
- He deals cards for a game of picture rummy.
- He gives each friend one of his toys with which to play.
- He divides his blocks so each child may build a house.

He learns that whole things can be divided into parts:

- One cookie is broken in half.
- An orange is divided into several segments.
- A carrot or banana is sliced into parts.
- The contents of a bottle of soda pop is put into two or more cups.
- A large piece of paper is cut into small pieces.

The young egocentric child centers on the number of things he sees. Two-year-old Pete breaks up his graham cracker into small pieces. "I have more than you," he says to Tim who has one whole graham cracker also. Pete does not see that although he has more pieces of cracker he does not have more crackers. Ms. Moore shows Chris a whole apple. "How many apples do I have?" "One,"

Fig. 11-1 Whole sandwiches are cut into parts.

Fig. 11-2 The banana is cut into two parts.

says Chris. "Now watch," says Ms. Moore as she cuts the apple into two pieces. "How many apples do I have now?" "Two!" answers Chris. As the child enters concrete operations, he will see that a single apple is always a single apple even though it may be cut into parts.

Gradually the child is able to see that a whole is made up of parts. He also begins to see that parts may be the same (equal) in size and amount or different (unequal) in size and amount. He compares number and size (unit 8) and develops the idea of more, less, and same. The idea of more, less, and same learned by the young child is needed later to understand that fractions such as one-half, one-fourth, and so on refer to equal parts.

ASSESSMENT

The teacher should observe as the child works and plays whether he uses the words *part* and *whole*. She should note if he uses them correctly. She should note his actions:

- Does he try to divide items to be shared equally among his friends?

- Will he think of cutting or breaking something in smaller parts if there is not enough for everyone?
- Does he realize when a part of something is missing (such as the wheel of a toy truck, the arm of a doll, the handle of a cup)?

Formal interview questions would be like the following:

Sample Assessment Task   **Preoperational Ages 2-3**

Parts/Wholes (unit 11): Missing Parts

Have real things and/or pictures of things with parts missing. Some examples are listed:

Things:	A doll with a leg or arm missing A car with a wheel missing A cup with a handle broken off A chair with a leg gone A face with only one eye A house with no door
Pictures:	Mount pictures of common things on poster board. Parts can be cut off before mounting.

Show each thing or picture and ask, LOOK CAREFULLY. WHICH PART IS MISSING FROM THIS <u>name of object</u>?

Sample Assessment Task   **Preoperational Ages 4-5**

Parts/Wholes (unit 11): Parts of a Whole

Show the child a whole apple. HOW MANY APPLES DO I HAVE? After it is certain that the child knows there is one apple, cut the apple in half. HOW MANY APPLES DO I HAVE NOW?

Sample Assessment Task   **Preoperational Ages 5-6**

Parts/Wholes (unit 11): Parts of Sets

Have three small child dolls (such as Weebles or Fisher-Price peg dolls). Have a box of pennies or other small objects. I WANT TO GIVE EACH CHILD SOME PENNIES. SHOW ME HOW TO DO IT SO EACH CHILD WILL GET SOME. Note how the child approaches the problem. Does he have the idea of giving one at a time to each, or does he use some other method for making the set of pennies into three small sets.

Fig. 11-3 Now the banana is cut two more times to make four parts.

The newborn infant is not aware that all his body parts are part of him. His early explorations lead him to find out that his hand is connected via an arm to his shoulder and those toes he sees at a distance are hooked to his legs. As he explores objects, he learns that they have different parts also. As he begins to sort and move objects about, he learns about parts and wholes of sets.

The following are some examples of the young child's use of the part/whole idea:

- Two-year-old Pete has a hot dog on his plate. The hot dog is cut in six pieces. He gives two pieces to his father, two to his mother, and keeps two for himself.

- Three-year-old Jim is playing with some toy milk bottles. He says to Ms. Brown, "You take two like me."

- Three-year-old Kate is sitting on a stool in the kitchen. She sees three eggs boiling in a pan on the stove. She points as she looks at her mother, "One for you, one for me, and one for Dad."

- Tanya is slicing a carrot. "Look I have a whole bunch of carrots now."

- Juanita is lying on her cot at the beginning of nap time. She holds up her leg. "Mrs. Raymond, is this part of a woman?"

- Bob runs up to Mr. Brown. "Look I have a whole tangerine."

INFORMAL ACTIVITIES

Many times during the day the teacher can help children develop their understanding of parts and wholes. The teacher can use the words *part, whole, divide* and *half*.

- Today everyone gets *half* of an apple and *half* of a sandwich.

- Too bad, *part* of this game is missing.

- Take this basket of crackers and *divide* them up so everyone gets some.

- No, we won't cut the carrots up. Each child gets a *whole* carrot.

- Give John *half* the blocks so he can build too.

- We only have one apple left. Let's *divide* it up.

- Point to the *part* of the body when I say the name.

The child can be given tasks which require him to learn about parts and wholes. When the child is asked to pass something, to cut up

Fig. 11-4 A big piece of play dough is divided into smaller parts.

vegetables or fruit, or to share materials, he learns about parts and wholes.

STRUCTURED ACTIVITIES

The child can be given structured experiences in all three types of part/whole relationships. Activities can be done which help the child become aware of special parts of people, animals, and things. Other groups of activities involve dividing sets into smaller sets. The third type of activity gives the child experiences in dividing wholes into parts.

PARTS AND WHOLES:
————PARTS OF THINGS————

Objective: To learn the meaning of the term *part* as it refers to parts of objects, people, and animals

Materials: Objects or pictures of objects with parts missing

Activities:

1. The Broken Toys
   Show the child some broken toys or pictures of broken toys. WHAT'S MISSING FROM THESE TOYS? After the child tells what is missing from each toy bring out the missing parts (or pictures of missing parts): FIND THE MISSING PART THAT COMES WITH EACH TOY.

2. Who (or What) is Hiding?
   The basic game is to hide someone or something behind a screen so that only a part is showing. The child then guesses who or what is hidden. The following are some variations:

   a. Two or more children hide behind a divider screen, a door, or a chair. Only a foot or a hand of one is shown. The other children guess whose body part can be seen.

   b. The children are shown several objects. The objects are then placed behind a screen. A part of one is shown. The child must guess which thing the part belongs to. (To make the task harder, the parts can be shown without the children knowing first what the choices will be).

Fig. 11-5   Now there are two parts made into balls.

   c. Do the same type of activity using pictures:

   • Cut out magazine pictures and mount on cardboard (or draw your own). Cut a piece of construction paper to use to screen most of the picture. Say, LOOK AT THE PART THAT IS SHOWING. WHAT IS HIDDEN BEHIND THE PAPER?

   • Mount magazine pictures on construction paper. Cut out a hole in another piece of construction paper of the same size. Staple the piece of paper onto the one with the picture on it so that a part of the picture can be seen. Say, LOOK THROUGH THE HOLE. GUESS WHAT IS IN THE PICTURE UNDER THE COVER.

Follow-up: Play *What's Missing Lotto* game. (Childcraft) or *What's Missing? Parts & Wholes, Young Learners Puzzles* (Teaching Resources)

—PARTS AND WHOLES: DIVIDING SETS—

Objective: To give the child practice in dividing sets into parts (smaller sets)

Materials: Two or more small containers and some small objects such as pennies, dry beans, or buttons

Activity: Set out the containers (start with two and increase the number as the child is able to handle more). Put the pennies or other objects in a bowl next to the containers. DIVIDE THESE UP SO EACH CONTAINER HAS SOME. Note whether the child goes about the task in an organized way and if he tries to put the same number in each container.

Follow-up: Increase the number of smaller sets to be made. Use different types of containers and different objects.

──────PARTS AND WHOLES──────

Objective: To divide whole things into two or more parts

Materials: Real things or pictures of things that can be divided into parts by cutting, tearing, breaking, or pouring

Activities:

1. Have the children cut up fruits and vegetables for snack or lunch. Be sure the children are shown how to cut so as not to hurt themselves. Be sure also that they have a sharp knife so the job is not frustrating. Children with poor coordination can tear lettuce, break off orange slices, and cut the easier things such as string beans.

2. Give the child a piece of paper. Have him cut it or tear it. Then have him fit the pieces back together. Have him count how many parts he made.

3. Give the child a piece of play dough or clay. Have him cut it with a dull knife or tear it into pieces. How many parts did he make?

4. Use a set of plastic measuring cups and a larger container of water. Have the children guess how many of each of the smaller cups full of water will fill the one cup measure. Let each child try the one-fourth, one-third, and one-half cups and count how many of each of these cups will fill the one cup.

Follow-up: Purchase or make some more structured part/whole materials such as:

1. Fraction Pies: circular shapes available in rubber and magnetic versions.

2. Materials which picture two halves of a whole.
   Halves to Wholes (DLM)
   Match-ups (Childcraft and Playskool)

3. Puzzles which have a sequence of difficulty with the same picture cut into two, three, and more parts:
   Basic Cut Puzzles (DLM)
   Fruit and Animal Puzzles (Teaching Resources)

4. Dowley Doos take-apart transportation toys (Lauri)

EVALUATION

The adult should observe and note if the child shows increased use of part/whole words and more skills in his daily activities:

- Can he divide groups of things into smaller groups?

- Can he divide wholes into parts?

- Does he realize that objects, people, and animals have parts that are unique to each?

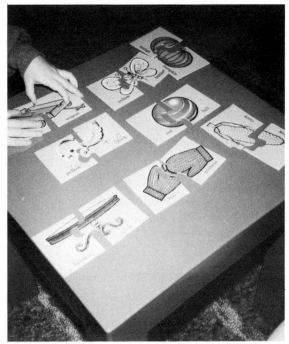

Fig. 11-6 Part/whole match-ups are a challenge to the young child.

SUMMARY

The young child learns about parts and wholes. The idea of parts and wholes is basic to what the child will learn later about fractions.

The child learns that things, people, and animals have parts. He learns that sets can be divided into parts (sets with smaller numbers of things.) He learns that whole things can be divided into smaller parts or pieces.

Experiences in working with parts and wholes help the young child move from preoperational centering to the concrete view and to understanding that the whole is no more than the sum of all its parts.

SUGGESTED ACTIVITIES

- Visit a Montessori classroom. Examine the materials which teach the ideas of parts and wholes. List these materials and describe how they are used. Report your experience to the class.

- Assess a young child's level of development in part-whole relations. Prepare an instructional plan based on the results. Prepare the materials and teach the child. Evaluate the results. Did the child learn from the instruction?

- Check through the Activities File. Be sure it is up to date with activities for each unit topic.

- Try out some of the structured activities described in this unit with children of different ages.

- If the school budget allocated twenty dollars for part/whole materials, decide on what would be top priority purchases.

REVIEW

A. Select the items in Column II which apply to the items in Column I.

Column I	Column II
1. Tina shares her cupcake.	a. Things, people, and animals have parts
2. Pieces of a monkey puzzle	
3. Tom gives Joe some of his crayons.	b. Sets can be divided into parts
4. Jim tears a piece of paper.	
5. Sue takes half the lump of dough.	c. Whole things can be divided into smaller parts or pieces
6. Chris puts the red blocks in one pile and the blue blocks in another.	
7. A doll's arm	
8. Stephanie takes the screwdriver from the tool kit.	
9. Orange section	
10. Piano keys	

B. Answer each of the following.

1. Briefly describe how a child first learns that wholes have parts.

2. What are the three types of part/whole relationships?

3. Describe two informal part/whole relationship activities.

4. Describe two structured part/whole relationship activities.

5. Describe one part/whole assessment task.

6. Describe how to evaluate a child's knowledge of parts and wholes through observation.

unit 12 the language of math

OBJECTIVES

After studying this unit, the student should be able to

- Explain two ways to describe a child's understanding of math words

- List the math words used in units five through eleven

- List math words used for ordering and measurement

What the child does and what the child says tell the teacher what the child knows about math. The older the child gets, the more important the math words he can use become. The child's language system is usually well developed by age four. That is, by age four his sentences are much the same as an adult's. He is at a point where the size of his vocabulary (the number of different words he can use) grows very fast.

The adult observes what the child does from infancy through age two and looks for the first understanding and the use of words. Between two and four the child starts to put more words together into longer sentences. He also learns more words and what they mean.

Fig. 12-1 *"One, two, three, four — I can dial numbers."*

In assessing the young child's development in math, questions are used. Which is the big ball? Which is the circle? The child's understanding of words is checked by having him respond with the right action:

- Point to the big ball.

- Find two chips.

- Show me the picture in which the boy is on the chair.

The above tasks do not require the child to say any words. He needs only point, touch, or pick up something. Once the child demonstrates his understanding of math words by using gestures or other nonverbal answers, he can move on to questions he must answer with one or more words. The child can be asked the same questions as above in a way that makes him speak:

- (The child is shown two balls, one big and one small.) "What is different about these balls?"

- (The child is shown a set of objects.) "How many are there in this set?"

- (The child is shown a picture of a boy sitting on a chair.) "Where is the boy?"

The child learns many math words as he goes about his daily activities. It has been found that by the time a child starts kindergarten, he uses many math words he has learned in a naturalistic way. Examples have

Fig. 12-2 "Are there *more short* pencils or *more long* pencils?"

been included in each of the previous units (five through eleven). The child uses both comments and questions. Comments would be like the following:

- Mom, I want *two* pieces of cheese.
- I have a *bunch* of bird seed.
- Mr. Brown, this chair is *small*.
- *Yesterday* we went to the zoo.
- The string is *long*.
- This is the *same* as this.
- The foot fits *in* the shoe.
- This cracker is a *square* shape.

Questions would be like these:

- How *old* is he?
- *When* is Christmas?
- *When* will I grow as *big* as you?
- *How many* are coming for dinner?
- Who has *more*?
- What *time* is my TV program?
- Is this a school *day,* or is it *Saturday?*

The answers that the child gets to these questions can help increase the number of math words he knows and can use.

The teacher needs to be aware of using math words during free play, lunch, and other times when a structured lesson in math is not being done. She should also note which words the child uses during free times.

The teacher should encourage the child to use math words even though he may not use them in an accurate, adult way. For example:

- I can count — one, two, three, five, ten.
- Aunt Helen is coming after my last nap. (Indicates future time)
- I will measure my paper. (Holds ruler against the edge of the paper.)
- Last night Grandpa was here. (Actually several days ago)
- I'm six years old. (Really two-years-old)
- I have a million dollars. (Has a handful of play money)

Adults should accept the child's use of the words and use the words correctly themselves. Soon the child will develop a higher level use of words as he is able to grasp higher level ideas. For the two- or three-year-old, any group of things more than two or three may

Fig. 12-3 "Hold up *two* fingers."

be called a *bunch*. Instead of using *big* and *little* the child may use family words: this is the mommy block and this is the baby block. Time (unit 15) is one idea that takes a long time to grasp. A young child may use the same word to mean different time periods. The following examples were said by a three-year-old:

- *Last night* we went to the beach (meaning last summer).

- *Last night* I played with Chris (meaning yesterday).

- *Last night* I went to Kenny's house (meaning three weeks ago).

For this child, *last night* means any time in the past. One by one he will learn that there are words that refer to times past such as last summer, yesterday, and three weeks ago.

Many math words have already been introduced and more will appear in the units to come. The prekindergarten child continually learns words. The following presents those math words which most children can use and understand by the time they enter kindergarten.

MATH WORDS

The following kinds of words have been used in units five through eleven:

- **Matching:** one, pair, more, each, some, group, bunch

- **Number and Counting:** zero, one, two, three, four, five, six, seven, eight, nine, ten, how many, count, group

- **Sets and Classifying:** sets; descriptive words for color, shape, size, materials, pattern, texture, function, association, class names, and common features; belong with; goes with; is used with; put with; the same

- **Comparing:** more, less, big, small, large, little, long, short, fat, skinny, heavy, light, fast, slow, cold, hot, thick, thin, wide, narrow, near, far, later, sooner, earlier, older, younger, newer, higher, lower, loud, soft (sound)

- **Shape:** circle, square, triangle, rectangle, oval, diamond, shape, round

- **Space:** *where* (on, off, on top of, over, under, in, out, into, out of, top, bottom, above, below, in front of, in back of, behind, beside, by, next to, between); *which way* (up, down, forward, backward, around, through, to, from, toward, away from, sideways, across); *distance* (near, far, close to, far from)

- **Parts and wholes:** part, whole, divide, share, pieces, some, half

Words which will be introduced later are also given:

- **Ordering:** first, second, third; big, bigger, biggest; few, fewer, fewest; large, larger, largest; little, littler, littlest; many, more, most; thick, thicker, thickest; thin, thinner, thinnest; last, next, then

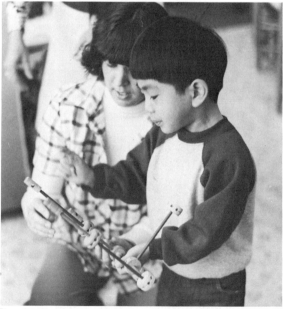

Fig. 12-4 "Your airplane can go *up high above* the houses."

- **Measurement of volume, length, weight, and temperature:** little, big, medium, tiny, large, size, tall, short, long, far, farther, closer, near, high, higher, thin, wide, deep, cup, pint, quart, gallon, ounces, milliliter, kiloliter, liter, foot, inch, kilometer, mile, meter, centimeter, narrow, measure, hot, cold, warm, cool, thermometer, temperature, ounces, pounds, grams, kilograms, milligrams

- **Measurement of time and sequence:** morning, afternoon, evening, night, day, soon, week, tomorrow, yesterday, early, late, a long time ago, once upon a time, minute, second, hour, new, old, already, Easter, Christmas, birthday, now, year, weekend, clock, calendar, watch, when, time, date, sometimes, then, before, present, soon, while, never, once, sometime, next, always, fast, slow, speed, Monday and other days of the week, January and other months of the year, winter, spring, summer, fall

- **Practical:** money, cash register, penny, dollar, buy, pay, change, cost, check, free, store, map

Fig. 12-5 *Today* **it is sunny outside.**

Words can be used before they are presented in a formal structured activity. The child who can talk can become familiar with words and even say them before he understands the idea they stand for.

SUMMARY

As the child learns the ideas and skills of math, he also learns many words. Math has a language that is basic to its activities. As the child learns math, he also increases his vocabulary and finds more uses for the same words.

SUGGESTED ACTIVITIES

- Visit a three-year-old, a four-year-old, and a five-year-old group of children. Observe for half an hour in each. Write down every child and adult math word used. Compare the three age groups and teachers for number and variety of words used.

- Make the math language assessment task materials described in the Appendix. Try them out on some four- and five-year-old children. Compare the number and variety of math words used. If others in your class do this activity, compare the results.

- Make a vocabulary list for your Activities File.

REVIEW

A. Briefly answer each of the following.

1. Explain two ways to describe a child's understanding of math words.

2. List ten math words used in units 5-11.

3. List ten math words used for ordering and measuring.

B. Complete the Math Word Puzzle.

Down	*Across*
1. Measurement word	1. Part/whole word
2. Part/whole word	2. Classifying word
3. Comparing word	3. Comparing word
4. Shape word	4. Classifying word
5. Measurement of height	5. Measurement of time word
6. Number word	6. Ordering
7. Matching word used for things in a group	7. Space word

Section 3 Using the Basics

unit 13 ordering

OBJECTIVES

After studying this unit, the student should be able to

- Define ordering
- List and describe the four basic types of ordering activities
- Do informal and structured ordering activities with young children
- Assess and evaluate a child's ability to order

Ordering is a higher level of comparing (unit 8). Ordering involves comparing more than two things or more than two sets. It also involves placing things in a sequence from first to last. In Piaget's terms, ordering is called *seriation.*

Ordering starts to develop in the sensori-motor stage. Before the age of two, the child likes to work with nesting toys. Nesting toys are items of the same shape but of varying sizes so that each one fits into the larger ones. If put into each other in order by size, they will all fit in one stack.

An early way of ordering is to place a pattern in one-to-one correspondence with a model as in figure 13-2 (A and B), page 98, This gives the child the idea of ordering. Next, he learns to place things in ordered rows on the basis of length, width, height, and size. At first the child can think of only two things at one time. When ordering by length, he places sticks in a sequence such as shown in figure 13-2(C). As he develops and practices, he will be able to use the whole sequence at once and place the sticks as in figure 13-2(D). Figures 13-3, 13-4, and 13-5 show a child as he goes through this process. As the child

develops further, he becomes able to order by other characteristics, such as color shades (dark to light), texture (rough to smooth), and sound (loud to soft).

Once the child can place one set of things in order, he can go on to double seriation. For double seriation he must put two groups of things in order, such as the pictures of girls with umbrellas as can be seen in figure 13-6, page 99. This is a use of matching, or one-to-one correspondence (unit 5).

Sets of things can also be put in order by the number of things in each set. By ordering

Fig. 13-1 This eighteen-month-old girl is challenged by nesting toys.

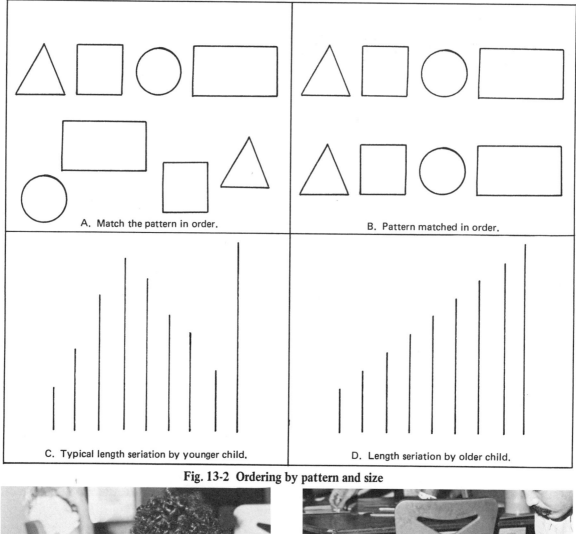

A. Match the pattern in order.

B. Pattern matched in order.

C. Typical length seriation by younger child.

D. Length seriation by older child.

Fig. 13-2 Ordering by pattern and size

Fig. 13-3 The child first explores the material.

Fig. 13-4 With teacher help, the child begins to learn about ordering.

Fig. 13-5 The child uses his senses to check what he has done.

Fig. 13-6 Which umbrella belongs to which girl?

sets, each with one more thing than the others, the child learns the concept (idea) of *one more than.* In figure 13-7, some cards with different numbers of dots are shown. In 13-7(A), the cards are mixed. In 13-7(B), the cards have been put in order so that each set has one more dot than the one before.

Ordering words are those such as next, last, biggest, smallest, thinnest, fattest, shortest, tallest, before, and after. Also included

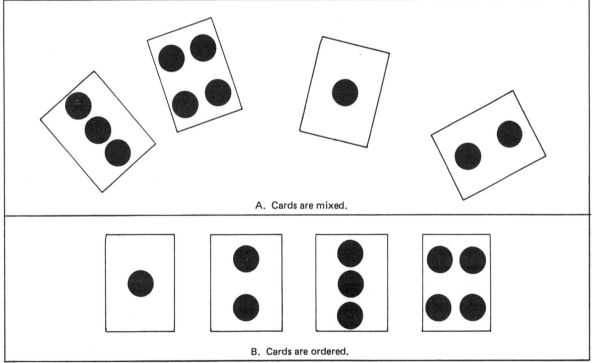

A. Cards are mixed.

B. Cards are ordered.

Fig. 13-7 Ordering Sets

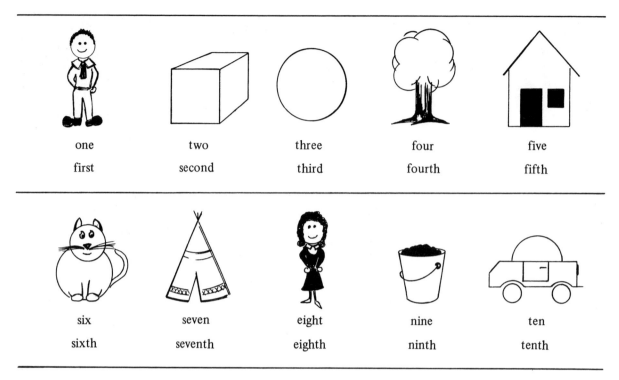

one · first
two · second
three · third
four · fourth
five · fifth

six · sixth
seven · seventh
eight · eighth
nine · ninth
ten · tenth

Fig. 13-8 Ordinal Numbers and Counting Numbers

are the ordinal numbers: first, second, third, fourth, and so on to the last thing. Ordinal relations as compared with counting are shown in figure 13-8.

ASSESSMENT

While the child plays, the teacher should note activities which might show the child is learning to order things. Notice how he uses nesting toys. Does he place them in each other so he has only one stack? Does he line them up in rows from largest to smallest? Does he use words such as first (I'm first.) and last (He's last.) on his own? In his dramatic play, does he go on train or plane rides where chairs are lined up for seats and each child has a place (first, second, last)?

Ask the child to order different numbers and kinds of items during individual interview tasks (see Appendix). In figure 13-9, the child has been asked to place the rectangles in order by width. The following are examples of two assessment tasks:

Sample Assessment Task   Preoperational Ages 4-5

Ordering (unit 13): Sequence/Ordinal Number

Present the child with four objects or pictures of objects which vary in height, width, or all size dimensions: FIND THE (TALLEST, BIGGEST, FATTEST) or (SHORTEST, SMALLEST, THINNEST). PUT THEM ALL IN A ROW FROM TALLEST TO SHORTEST (BIGGEST TO LITTLEST, FATTEST TO THINNEST). If the child accomplishes this task ask him, WHICH IS FIRST? WHICH IS LAST? WHICH IS SECOND? WHICH IS THIRD? WHICH IS FOURTH?

Sample Assessment Task   Transition: Preoperational to Concrete Operations Ages 5-7

Ordering (unit 13): Double Seriation

Have two sets of ten objects or pictures of objects such that there is one item in each set that is the right size for an item in the other set. The sets could be boys and baseball bats, girls and umbrellas, chairs and tables, bowls and spoons, cars and garages, and so on. Hats and heads with faces may be used. The heads are placed in front of the child: LINE THESE UP SO THAT THE SMALLEST HEAD IS FIRST AND THE

BIGGEST IS LAST. Help can be given, such as: FIND THE SMALLEST. OKAY, WHICH ONE COMES NEXT? AND NEXT?.... If the child is able to line them up correctly put out the hats: Say, FIND THE HAT THAT FITS EACH HEAD AND PUT IT ON THE HEAD.

NATURALISTIC ACTIVITIES

Just as the child's natural development leads him to sort things, it also leads him to put them in order. In fact after he sorts things out, he often then groups them in rows. He may pick out some blocks, all of one size and shape, and line them up in a row. Mother's measuring spoons and measuring cups are ideal for exploring size. As he pulls mixing bowls out of the kitchen cupboard, he checks to find out which ones fit into the others.

As speech increases, the child uses order words. "I want to be *first*." "This is the *last* one." "Daddy Bear has the *biggest* bowl." "I'll sit in the *middle*." As he starts to draw pictures, he often draws Moms, Dads, and children and places them in a row from smallest to largest.

INFORMAL ACTIVITIES

Informal teaching can go on quite often during the child's daily play and routine activities. The following are some examples:

• Eighteen-month-old Brad has a set of mixing bowls and measuring cups to play with on the kitchen floor. He puts the biggest bowl on his head. His mother smiles and says, "The *biggest* bowl fits on your head." He tries the smaller bowls, but they do not fit. Mom says, "The *middle sized* bowl and the *smallest* bowl don't fit, do they?" She sits down with him and picks up a measuring cup. "Look, here is the cup that is the *biggest*. These are *smaller*." She lines them up by size. "Can you find the *smallest* cup?" Brad proceeds to put the cups

Fig. 13-9 **Which rectangle is the widest?**

one in the other until they are in one stack. His mother smiles, "You have them all in *order*."

• Five-year-old George, four-year-old Richard, and three-year-old Jim come running across the yard to Mr. Brown. "You are all fast runners." "I was *first*," shouts George. "I was *second*," says Richard. "I was *third*," says Jim. George shouts, "You were *last*, Jim, 'cause you are the *littlest*." Jim looks mad. Mr. Brown says, "Jim was both *third* and *last*. It is true Jim is the *littlest* and the *youngest*. George is the *oldest*. Is he the *biggest*?" "No!" says Jim, "He's *middle size*."

• Mary has some small candies she is sharing with some friends and her teacher. "Mr. Brown, you and I get five because we are the *biggest*. Diana gets four because she's the *next* size. Pete gets three. Leroy gets two and Brad gets one. Michael doesn't get any 'cause he is a baby." "I see," says Mr. Brown, "You are dividing them so the *smallest* people get the least and the *biggest* get the most."

Fig. 13-10 The child learns with tactile material.

- Miss Collins tells the children, "You have to take turns on the swing. Tanya is *first* today."

These examples show how comments can be made which help the child see his own use of order words and activities. Many times in the course of the day, opportunities come up where children must take turns. These times can be used to the fullest for teaching order and ordinal number. Many kinds of materials can be put out for children which help them practice ordering. Some of these things are self-correcting as shown in figure 13-10.

STRUCTURED ACTIVITIES

Structured experiences with ordering ideas can be done with many kinds of materials. These materials can be purchased, or they can be made by the teacher. Things of different sizes are easy to find at home or school. Measuring cups and spoons, mixing bowls, pots and pans, shoes, gloves, and other items of clothing are easy to get in several different sizes. Paper and cardboard can be cut into different sizes and shapes. Paper towel rolls can be made into cylinders of graduated sizes. The artistic teacher can draw pictures of the same item in graduated sizes. The following are basic activities which can be done with many different kinds of objects and pictures.

——ORDERING: THE IDEA OF ORDER——

Objective: To help the child understand the idea of order and sequence

Materials: Large colored beads with a string for the teacher and each child

Activity: The beads are in a box or bowl where they can be reached by each child. Say, WATCH ME. I'M GOING TO MAKE A STRING OF BEADS. Start with three beads. Add more as each child learns to do each amount. Lay the string of three beads down where each child can see it: NOW YOU MAKE ONE LIKE MINE. WHICH KIND OF BEAD SHOULD YOU TAKE FIRST? When the first bead is on: WHICH ONE IS NEXT? When two are on: WHICH ONE IS NEXT?

Follow-up:

1. Make a string of beads. Pull it through a paper towel roll so that none of the beads can be seen. Say, I'M GOING TO HIDE THE BEADS IN THE TUNNEL. NOW I'M GOING TO PULL THEM OUT. WHICH ONE WILL COME OUT FIRST? NEXT? NEXT? and so on. Then pull the beads through and have the children check as each bead comes out.

2. Dye some macaroni with food coloring. Set up a pattern for a necklace. The children can string the macaroni to make their own necklaces in the same pattern.

ORDERING: DIFFERENT SIZES, ——————SAME SHAPE——————

Objective: To make comparisons of three or more items of the same shape and different sizes

Materials: Four to ten squares cut with sides one inch, one and one-quarter inch, one and one-half inch, and so on

Activity: Lay out the shapes: HERE ARE SOME SQUARES. STACK THEM UP SO THE BIGGEST IS ON THE BOTTOM. Mix the squares up again: NOW, PUT THEM IN A ROW STARTING WITH THE SMALLEST.

Follow-up: Do the same thing with other shapes and materials.

ORDERING: LENGTH

Objective: To make comparisons of three or more things of the same width but different lengths

Materials: Sticks, strips of paper, yarn, string, cuisinaire rods, drinking straws, or anything similar cut in different lengths such that each one is the same difference in length from the next one.

Activity: Put the sticks out in a mixed order: LINE THESE UP FROM SMALLEST TO LARGEST (LARGEST TO SMALLEST). Help if needed: WHICH ONE COMES NEXT? WHICH ONE OF THESE IS BIGGEST? IS THIS THE NEXT ONE?

Follow-up: Do this activity with many different kinds of materials.

ORDERING: DOUBLE SERIATION

Objective: To match one-to-one two ordered sets of the same number of items

Materials: Three Bears flannel board set or cutouts made by hand: a mother bear, father bear, baby bear, Goldilocks, the three bowls, three spoons, three chairs, and three beds

Activity: Tell the story. Use all the order words/ Biggest, middle sized, smallest, next. Follow-up with questions: WHICH IS THE BIGGEST BEAR? FIND THE BIGGEST BEAR'S BOWL (CHAIR, BED, SPOON). Use the same sequence with each character.

Follow-up: Let the children act out the story with the felt pieces or cutouts. Note if they use the order words, if they change their voices, and if they match each bear to the right bowl, spoon, chair, and bed.

ORDERING: SETS

Objective: Two order sets of one to five objects

Materials: Glue buttons on cards or draw dots on cards so there are five cards.

Activity: Lay out the cards. Put the card with one button in front of the child: HOW MANY BUTTONS ON THIS CARD? Child answers. Say, YES, THERE IS ONE BUTTON. FIND THE CARD WITH ONE MORE BUTTON. If child picks out the card with two: GOOD, NOW FIND THE CARD WITH ONE MORE BUTTON. Keep on until all five are in line. Mix the cards up. Give the stack to the child. LINE THEM ALL UP BY YOURSELF. START WITH THE SMALLEST SET.

Follow-up: Repeat with other materials. Increase the number of sets as each child learns to recognize and count larger sets. Use loose buttons (chips, sticks, or coins) and have the child count out his own sets. Each set can be put in a small container or on a small piece of paper.

ORDER: ORDINAL NUMBERS

Objective: To learn the ordinal numbers *first, second, third,* and *fourth* (The child should be able to count easily to four before he does these activities.)

Materials: four balls or beanbags, four common objects, four chairs

Activities:

1. Games requiring that turns be taken can be used. Just keep in mind that young children cannot wait very long. Limit the group to four children and keep the game moving fast. For example, give each of the four children one beanbag or one ball. Say, HOW MANY BAGS ARE THERE? LET'S COUNT. ONE, TWO, THREE, FOUR. CAN I CATCH THEM ALL AT THE SAME TIME? NO, I CAN'T. YOU WILL HAVE TO TAKE TURNS: YOU ARE FIRST, YOU ARE SECOND, YOU ARE THIRD, AND YOU ARE FOURTH. Have each child say his number, "I am (first, second, third, and fourth)." OKAY, FIRST, THROW YOURS. (throw it back) SECOND, THROW YOURS. (throw it back) After each has had his turn,

have them all do it again. This time have them tell you their ordinal number name.

2. Line up four objects. Say, THIS ONE IS FIRST, THIS ONE IS SECOND, THIS ONE IS THIRD, THIS ONE IS FOURTH. Ask the children: POINT TO THE (FOURTH, FIRST, THIRD, SECOND).

3. Line up four chairs. WE ARE GOING TO PLAY BUS (PLANE, TRAIN) Name a child, _____ YOU GET IN THE THIRD SEAT. Fill the seats. Go on a pretend trip. NOW WE WILL GET OFF. SECOND SEAT GET OFF. FIRST SEAT GET OFF. FOURTH SEAT GET OFF. THIRD SEAT GET OFF.

Follow-up: Make up some games which use the same basic ideas. As each child knows first through fourth, add fifth, then sixth, and so on.

EVALUATION

The teacher should note whether the child uses more ordering words and activities during his play and routine activities. Without disrupting the child, she should ask questions, and make comments and suggestions.

- Who is the biggest? (the smallest?)

SUGGESTED ACTIVITIES

- Observe children at play in school. Note those activities which show the child is learning to order things. Share this experience with your class.

- Add ordering activities to your Activities File.

- Make up some ordering assessment tasks. Try out some with young children. What did the children do? Share this experience with the class.

- While working with children, plan to do at least one informal and one structured ordering experience. Share what happened with the class.

- Look through two or three educational materials catalogues. Pick out ten materials which you feel would be the best for ordering activities.

- Make three sets of ordering materials that children can try during free play. Try them out with two different age groups of children.

- (As the children put their shoes on after their nap) Who has the longest shoes? (the shortest shoes?)

- Who came in the door first today?

- Run fast. See who can get to the other side of the gym first.

- (The children are playing train) Well, who is in the last seat? He must be the caboose. Who is in the first seat? He must be the engineer.

- Everyone can't get a drink at the same time. Line up with the shortest person first.

The assessment tasks in the Appendix may be used for formal evaluation interviews.

SUMMARY

When more than two things are compared, the process is called ordering, or seriation. There are four basic types of ordering activities. The first is to put things in sequence by size. The second is to make a one-to-one match between two sets of related things. The third is to place sets of different numbers of things in order from the least to the most. The last is ordinal number. Ordinal numbers are first, second, third, and so on.

REVIEW

A.   Indicate the statements which apply to ordering.

1.   Ordering is the same as counting.

2.   Ordering is called seriation.

3.   Ordering begins in the concrete operational stage of development.

4.   Ordering involves comparing attributes of things.

5.   Ordering involves placing things in sequence from first to last.

6.   Ordering involves comparing more than two things.

7.   Ordering may involve one-to-one correspondence.

8.   Ordering begins with putting two groups of things in order.

9.   Ordering is a higher level of comparing.

B.   Complete each of the following.

1.   Define ordering.

2.   What is another name for ordering?

3.   What are the four basic types of ordering activities?

C.   Match each item in Column II with the correct activity in Column I.

Column I

1.   Child lines up buttons on cards; one button on the first to five on the last.

2.   Child says, "I'm next."

3.   Child lines up straws of various sizes.

4.   Child stacks nesting cups.

5.   Child strings beads of different sizes according to pattern.

6.   Child puts one poker chip with one poker chip.

7.   Child puts the different sized flowers in the different pots of the right size.

8.   Teacher says "Line up with the tallest child first."

9.   Child says, "You're first."

10. Child builds a tower with a large block at the bottom, the next block is smaller, and the smallest on top.

Column II

a. Size sequence

b. One-to-one comparison

c. Ordering sets

d. Ordinal words or numbers

e. The basic idea of order

f. Double seriation

unit 14 measurement:
volume, weight, length, and temperature

OBJECTIVES

After studying this unit, the student should be able to

- Explain how measurement develops in five stages
- Assess and evaluate the measurement skills of a young child
- Do informal and structured measurement with young children

Measurement is one of the most useful math skills. *Measurement* involves assigning a number to things so they can be compared on the same attributes. Numbers can be assigned to attributes such as volume, weight, length, and temperature. For example, the child drinks *one cup* of milk. Numbers can also be given to time measurement. However, time is not an attribute of things and so is presented separately (unit 15). Standard units such as pints, quarts, liters, yards, meters, pounds, grams, and degrees tell us exactly how much (*volume*); how heavy (*weight*); how long, wide, or deep (*length*); and how hot or cold (*temperature*). A number is put with a standard unit to let a comparison be made. Two quarts is more than one quart, two pounds is lighter than three pounds, one meter is shorter than four meters, and 30° is colder than 80°.

STAGES OF DEVELOPMENT

The concept of measurement develops through five stages as outlined in figure 14-1. The first stage is a play stage. The child imitates older children and adults. He plays at measuring with rulers, measuring cups, measuring spoons, and scales as he sees others do. He pours sand, water, rice, beans, and peas from one container to another as he explores the properties of volume. He lifts and moves things as he learns about weight. He notes that those who are bigger than he is can do many more activities and has his first concept of length (height). He finds that his short arms cannot always reach what he wants them to reach (length). He finds that he has a preference for cold or hot food and cold or hot bath water. He begins to learn about temperature. This first stage begins at birth

Piagetian Stage	Age	Measurement Stage
Sensorimotor and Preoperational	Birth to Age Seven Years	1. Plays and imitates 2. Makes comparisons
Transition: Preoperational to Concrete Operations	Five to Seven Years	3. Uses arbitrary units
Concrete Operations	Six Years Or Older	4. Sees need for standard units 5. Uses standard units

Fig. 14-1 Stages in the Development of the Concept of Measurement

and continues through the sensorimotor period into the preoperational period.

The second stage in the development of the concept of measurement is the one of making comparisons (unit 8). This is well under way by the preoperational stage. The child is always comparing: bigger-smaller, heavier-lighter, longer-shorter, and hotter-colder.

The third stage which comes at the end of the preoperational period and at the beginning of concrete operations is one in which the child learns to use what are called arbitrary units. That is, anything the child has can be used as a unit of measure. He will try to find out how many coffee cups of sand will fill a quart milk carton. The volume of the coffee cup is the arbitrary unit. He will find out how many toothpicks long his foot is. The length of the toothpick is the arbitrary unit. As he goes through the stage of using arbitrary units, he learns concepts he will need to understand standard units.

When the child enters the period of concrete operations, he can begin to see the need for standard units. He can see that to communicate with someone else in a way the other person will understand, he must use the same units the other person uses. For example, the child says that his paper is nine thumbs wide. Another person cannot find another piece of the same width unless the child and the thumb are there to measure it. But, if he says his paper is eight and one-half inches wide, another person will know exactly the width of the paper. In this case, the thumb is an arbitrary unit, and the inch is a standard unit. The same is true for other units. To say "two liters of water" tells more than saying "two pitchers of water." "The box weighs three kilos" tells more than "the box weighs the same as four buckets of sand."

The last stage in the development of the concept of measurement begins in the concrete

Fig. 14-2 Through play, the child learns to measure.

operations period. In this last stage, the child begins to use and understand the standard units of measurement such as inches, meters, pints, liters, grams, and degrees.

It can be seen that most of the children to whom this book is directed will be in the first two stages. A few will be in the third stage. Probably none will be in stages four or five.

HOW THE YOUNG CHILD THINKS ABOUT MEASUREMENT

To find out why standard units are not understood by young children in the sensori-motor and preoperational stages, Piaget must be reviewed. Remember from the first unit that the young child is fooled by appearances.

Fig. 14-3 Arbitrary units are used to measure.

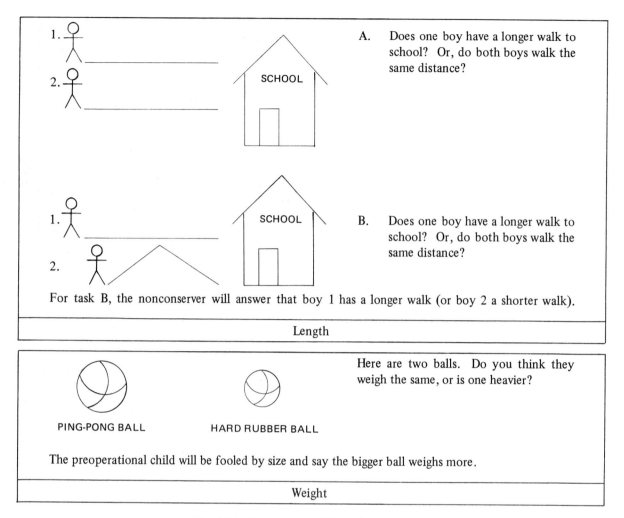

A. Does one boy have a longer walk to school? Or, do both boys walk the same distance?

SCHOOL

B. Does one boy have a longer walk to school? Or, do both boys walk the same distance?

SCHOOL

For task B, the nonconserver will answer that boy 1 has a longer walk (or boy 2 a shorter walk).

Length

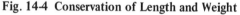

Here are two balls. Do you think they weigh the same, or is one heavier?

PING-PONG BALL   HARD RUBBER BALL

The preoperational child will be fooled by size and say the bigger ball weighs more.

Weight

Fig. 14-4 Conservation of Length and Weight

He believes what he sees before him. He does not keep old pictures in mind as he will do later. He is not yet able to conserve (or save) the first way something looks when its appearance is changed. When the ball of clay is made into a snake, he thinks the volume (the amount of clay) has changed because it looks smaller to him. When the water is poured into a different shaped container, he thinks there is more or less — depending on the height of the glass.

Two more examples are shown in figure 14-4. In the first task, the child is fooled when a crooked road is compared with a straight road. The straight road looks longer (conservation of length). In the second task,

size is dominant over material and the child guesses that the table tennis ball weighs more than the hard rubber ball. He thinks that since the table tennis ball is larger than the hard rubber ball, it must be heavier.

The young child becomes familiar with the words of measurement and learns which attributes can be measured. He learns mainly through observing older children and adults as they measure. He does not need to be taught the standard units of measurement in a formal way. The young child needs to gain a feeling that things differ on the basis of "more" and "less" of some attributes. He gains this feeling mostly through his own observations and firsthand experimental experiences.

ASSESSMENT

To assess measurement skills in the young child, the teacher observes. She notes whether the child uses the term *measurement* in the adult way. She notes if he uses adult measuring tools in his play as he sees adults use them. She looks for the following kinds of incidents:

- Mary is playing in the sandbox. She pours sand from an old bent measuring cup into a bucket and stirs it with a sand shovel. "I'm measuring the flour for my cake. I need three cups of flour and two cups of sugar."

- Juanita is seated on a small chair. Kate kneels in front of her. Juanita has her right shoe off. Kate puts Juanita's foot on a ruler. "I am measuring your foot for your new shoes."

- The children have a play grocery store. George puts some plastic fruit on the toy scale. "Ten pounds here."

- Tim is the doctor and Bob is his patient. Tim takes an imaginary thermometer from Bob's mouth. "You have a hot fever."

Individual interviews for the preoperational child may be found in unit 8. For the child who is near concrete operations (past five years of age), the conservation tasks in the Appendix and in unit 1 may be used.

NATURALISTIC ACTIVITIES

The young child's ideas about measurement come for the most part from his natural everyday experiences. The examples in the assessment section of this unit show that each child shows in his play activities some idea he has learned in his daily life. Mary has seen and may have helped someone make a cake. Kate has been to the shoe store and knows the clerk must measure the feet before he brings a pair of shoes to try on the customer. George has seen the grocer weigh fruit. Tim knows that a thermometer tells how "hot" a fever is. The observant young child picks up these ideas on his own without being told specifically that they are important.

The child uses his play activities to practice what he has seen adults do. He also uses play materials to learn ideas through trial and error and experimentation. Water, sand, dirt, mud, rice, and beans teach the child about volume. As he pours these substances from one container to another, he learns about *how much*, or amount. The child can use containers of many sizes and shapes: buckets, cups, plastic bottles, dishes, bowls, and coffee cans. Shovels, spoons, strainers, and funnels can also be used with these materials. When playing with water, the child can also learn about weight if he has some small objects like sponges, rocks, corks, small pieces of wood, and marbles which may float or sink. Any time a child tries to put something in a box, envelope, glass, or any other container, he learns something about volume.

Fig. 14-5 These girls learn about volume.

The child can begin to learn the idea of linear measure (length, width, height) and area in his play. The unit blocks which are usually found in the early childhood classroom help the child learn the idea of units. He will soon learn that each block is a unit of another block. Two, four, or eight of the small blocks are the same length when placed end to end as one of the longest blocks. As he builds enclosures (houses, garages, farmyards, etc.), he is forced to pick his blocks so that each side is the same length as the one across from it.

The child learns about weight and balance on the teeter-totter. He soon learns that it takes two to go up and down. He also learns that it works best when the two are near the same weight and are the same distance from the middle.

The child makes many contacts with temperature. He learns that his soup is hot, warm, and then as it sits out, turns cold. He likes cold milk and hot cocoa. He learns that the air may be hot or cold. If the air is hot, he may wear just shorts or a bathing suit. If the air is cold, he will need a coat, hat, and mittens.

INFORMAL ACTIVITIES

The young child learns about measurement through the kinds of experiences just described. During these activities there are many opportunities for informal teaching. One job for the adult as the child plays is to help him by pointing out properties of materials which the child may not be able to find on his own. For instance, if a child says he must have all the long blocks to make his house large enough, the teacher can show him how several small blocks can do the same job. She can show the child how to measure how much string will fit around a box before he cuts off a piece to use.

The teacher can also take these opportunities to use measurement words such as the names of units of measurement and the words listed in unit 12. She can also pose problems for the child:

- How can we find out if we have enough apple juice for everyone?

- How can we find out how many paper cups of milk can be poured from a gallon container?

- How can we find out if someone has a high fever?

- How can we find out without going outside if we need to wear a sweater or coat?

- How can we find out who is the tallest boy in the class? The heaviest child?

- How many of these placemats will fit around the table?

- Who lives the longest distance from school?

It is up to the teacher to be alert for opportunities to help the child in his search for the ideas which are part of measurement.

STRUCTURED ACTIVITIES

The young child learns most of his basic measurement ideas through his play and home activities that come through the natural routines of the day. He gains a feeling for the need for measurement and learns the language of measurement. Structured activities must be chosen with care. They should make use of the child's senses. They should be related to what is familiar to the child and expand what he already knows. They should pose problems which will show him the need for measurement. They should give the child a chance to use measurement words to explain his solution to the problem. The following are examples of these kinds of experiences.

Fig. 14-6 Experimenting with water and containers helps children learn about volume.

————MEASUREMENT: VOLUME————

Objectives:

- To learn the characteristics of volume
- To see that volume can be measured
- To learn measurement words used to tell about volume (more, less, too big, too little, the same)

Materials:

- Sandbox (indoors and/or out), water table (or sink or plastic dishpans)
- Many containers of different sizes: bottles, cups, bowls, milk cartons, cans (with smooth edges), boxes (for dry materials)
- Spoons, scoops, funnels, strainers, beaters
- Water, sand, rice, beans, peas, or anything else that can be poured

Activities:

1. Allow plenty of time for experimentation with these kinds of materials during free playtime.

2. Have several containers of different kinds and sizes. Fill one with water (or sand or rice or peas or beans). Pick out another container. Ask the children: IF I POUR THIS WATER FROM THIS BOTTLE INTO THIS OTHER BOTTLE, WILL THE SECOND BOTTLE HOLD ALL THE WATER? After each child has made his prediction, pour the water into the second container. Ask a child to tell what he saw happen. Continue with several containers. Have the children line them up from the one that holds the most to the one that holds the least.

3. Pick out one standard container (coffee cup, paper cup, measuring cup, tin can). Have one or more larger containers. Say, IF I WANT TO FILL THE BIG BOWL WITH SAND AND USE THIS PAPER CUP, HOW MANY TIMES WILL I HAVE TO FILL THE PAPER CUP AND POUR SAND INTO THE BOWL? Write down the children's predictions. Let each child have a turn to fill the cup and pour sand into the bowl. Record by making slash marks how many cups of sand are poured. Have the children count the number of marks when the bowl is full. Compare this amount with what the children thought the amount would be. This can be done with many different sets of containers.

Follow-up: Do the same types of activities using different sizes of containers and common objects. For example, have a doll and three different size boxes. Have the children decide which box the doll will fit into.

————MEASUREMENT: WEIGHT————

Objectives:

- To learn firsthand the characteristics of weight
- To learn that weight and size are different attributes (big things may have less weight than small things)
- To learn that light and heavy are relative ideas

Materials:

- Things in the classroom
- A teeter-totter, a board and a block, a simple pan balance

- Sand, sugar, salt, flour, sawdust, peas, beans, rice

- A ball collection with balls of different sizes and materials: ball bearings, table tennis, golf, solid rubber, foam rubber, styrofoam, balsam wood, cotton, balloons

Activities:

1. Have the child name things in the room that he can lift and things he cannot lift. Which things can he not lift because of size? Which because of weight? Compare things such as a stapler and a large paper bag (small and heavy and large and light). Have the children line things up from heaviest to lightest.

2. Have the children experiment with the teeter-totter. How many children does it take to balance the teacher? Make a balance with a block and a board. Have the child experiment with different things to see which will make the board balance.

3. A fixed position pan balance can be used for firsthand experiences with all types of things:

   a. The child can try balancing small objects such as paper clips, hair clips, bobby pins, coins, toothpicks, cotton balls, and so on in the pans.

   b. Take the collection of balls. Pick out a pair. Have the child predict which is heavier (lighter). Let him put one in each pan to check his prediction.

   c. Put one substance such as salt in one pan. Have the child fill the other pan with flour until the pans balance. IS THE AMOUNT (VOLUME) OF FLOUR AND SALT THE SAME?

   d. Have equal amounts of two different substances such as sand and sawdust in the balance pans. DO THE PANS BALANCE?

Follow-up: Make some play dough with the children. Have them measure out one part flour and one part salt. Mix in some powder tempera. Add water until the mixture is pliable but not too sticky. See unit 17 for cooking ideas.

MEASUREMENT: LENGTH AND HEIGHT

Objectives:

- To learn firsthand the concepts of length and height
- To help the child learn the use of arbitrary units

Materials:

- The children themselves
- Things in the room
- Balls of string and yarn, scissors, construction paper, magic marker, beans (chips, pennies, or other small counters), pencils, toothpicks, popsicle sticks, unit blocks.

Activities:

1. Present the child with problems where he must pick out something of a certain length. For example, a dog must be tied to a post. Have a picture of the dog and the post. Have several lengths of string. Have the child find out which string is the right length. Say: WHICH ROPE WILL REACH FROM THE RING TO THE DOG'S COLLAR?

2. LOOK AROUND THE ROOM. WHICH THINGS ARE CLOSE? WHICH THINGS ARE FAR AWAY?

3. Have several children line up. Have a child point out which is the tallest, the shortest. Have the children line up from tallest to shortest. The child can draw pictures of friends and family in a row from shortest to tallest.

4. Draw lines on construction paper. HOW MANY BEANS (CHIPS, TOOTHPICKS OR OTHER SMALL THINGS) WILL FIT ON EACH LINE? WHICH LINE HAS MORE BEANS? WHICH LINE IS LONGEST? Gradually use paper with more than two lines.

5. Put a piece of construction paper on the wall from the floor up to about five feet. Have each child stand next to the paper. Mark his height, write his name by his height. Check each child's height each month. Note how much each child grows over the year.

6. Have an arbitrary unit such as a pencil, a toothpick, a stick, a long block, or a piece of yarn or string. Have the child measure things in the room to see how many units long, wide, or tall the things are.

Follow-up: Keep the height chart out so the children can look at it and talk about their heights.

——MEASUREMENT: TEMPERATURE——

Objectives:

- To give the child firsthand experiences which will help him learn that temperature is the relative measure of heat

- To learn that the thermometer is used to measure temperature

- To experience hot, warm, and cold as related to things, to weather, and to the seasons of the year

Materials: Ice cubes, hot plate, teakettle or pan, pictures of the four seasons, posterboard, magic markers, scissors, glue, construction paper, old magazines with pictures, real thermometers (body, inside, and outside)

Activities:

1. Have the children decide whether selected things in the environment are hot, cold, or warm: ice and boiling water, the hot and cold water taps, the radiators, the glass in the windows, their skin, for example.

2. Show pictures of summer, fall, winter and spring. Discuss the usual temperatures in each season. What is the usual weather? What kinds of clothes are worn? Make a cardboard thermometer. At the top put a child in heavy winter clothes, underneath put a child in a light coat or jacket, then a child in a sweater, then one in short sleeves, then one in a bathing suit. Each day discuss the outside temperature relative to what was worn to school.

3. Give the children scissors and old magazines. Have them find and cut out pictures of hot things and cold things. Have them glue the hot things on one piece of poster board and the cold things on another.

4. Show the children three thermometers: one for body temperature, one for room temperature, and one for outdoor use. Discuss when and where each is used.

Follow-up: Each day the outside temperature can be discussed and recorded in some way (such as in the second listed activity or on a graph as discussed in unit 16).

EVALUATION

The adult should note the children's response to the activities given them. She should observe them as they try out the materials and note their comments. She must also observe whether they are able to solve everyday problems that come up by using informal measurements such as comparisons. Use the individual interviews in the Appendix.

SUMMARY

Measurement develops through five stages. The young child is in the early stages: play and imitation and comparing. He learns about measurement mainly through naturalistic and informal experiences.

Fig. 14-7 Grandmother checks toddler's weight.

SUGGESTED ACTIVITIES

- Observe preschoolers in group play. Write down any examples of measurement activities that occur. Which stage of measurement did each represent?

- With a small group of classmates, discuss some ways a home or school environment can encourage children to use measurement skills.

- Plan some structured measurement activities. Try to use one with a small group of preschoolers. Share with the class what happened.

- Look through a toy catalog for measurement materials that can be purchased for young children.

- Make a balance that could be used with young children.

- Add at least five structured measurement activities to the Math Activities File.

REVIEW

A. Put the following in order beginning with the first stage of measurement.

    a. sees need for standard units   d. uses standard units

    b. arbitrary units used   e. makes comparisons

    c. play and imitation

B. List and describe the measurement stages.

C. Five levels of measurement have been discussed. After looking at the following statements, identify the measurement level which best fits each situation.

    1. Tommy says, "My truck is bigger than yours."

    2. Mary checks the outdoor thermometer and says, "It's 72 degrees out today.

    3. Sara, Vera, and Joe pour sand in and out of various containers.

    4. "I weigh fifty pounds. How much do you weigh?"

    5. "Dad, two of my shoes are the same length as one of yours."

D. Discuss how you could assess and evaluate a child's measurement skills.

unit 15 measurement: time

OBJECTIVES

After studying this unit, the student should be able to

- Describe what is meant by time sequence
- Describe what is meant by time duration
- Explain the three kinds of time
- Do informal and structured time measurement activities with young children

There are two sides to the concept of time. There is sequence and there is duration. Sequence of time has to do with the order of events. It is related to the ideas about ordering presented in unit 13. While the child learns to sequence things in patterns, he also learns to sequence events. He learns small, middle sized, and large beads go in order for a pattern sequence. He gets up, washes his face, brushes his teeth, dresses, and eats breakfast for a time sequence. Duration of time has to do with how long an event takes (seconds, minutes, hours, days, a short time, a long time).

KINDS OF TIME

There are three kinds of time a child has to learn. Time is a hard measure to learn. The child cannot see it and touch it as he can weight, volume, length, and temperature. There are fewer clues to help the child. The young child relates time to three things: personal experience, social activity, and cultural time. In his *personal experience*, the child has his own past, present, and future. The past is often referred to as "When I was a baby." "Last night" may mean any time before right now. The future may be "After my night nap" or "When I am big." The young child has difficulty with the idea that there was a time when mother and dad were little and he was not yet born.

Time in terms of *social activity* is a little easier to learn and makes more sense to the young child. The young child tends to be a slave to order and routine. A change of schedule can be very upsetting. This is because time for him is a sequence of predictable events. He can count on his morning activities being the same each day when he wakes up. Once he gets to school, he learns that there is order there too: first he takes off his coat and hangs it up, next he is greeted by his teacher, then he goes to the big playroom to play, and so on through the day.

A third kind of time is *cultural time*. It is the time that is fixed by clocks and calendars.

Fig. 15-1 **Charlene works with beads and learns pattern sequence.**

115

It is the time that is the same for everyone. It is a kind of time that the child probably does not really understand until he is in the concrete operations period. He can, however, learn the language (seconds, minutes, days, months, etc.) and the names of the timekeepers (clock, watch, calendar). He can also learn to recognize a timekeeper when he sees one.

LANGUAGE OF TIME

To learn time is as dependent on language as any part of math. Time and sequence words are listed in unit 12. They are listed again in this unit for easy reference:

- **General words:** time, age

- **Specific words:** morning, afternoon, evening, night, day, noon

- **Relational words:** soon, tomorrow, yesterday, early, late, a long time ago, once upon a time, new, old, now, when, sometimes, then, before, present, soon, while, never, once, next, always, fast, slow, speed, first, second, third, and so on.

- **Specific duration words:** Clock and watch (minutes, seconds, hours) Calendar (date, days of the week names, names of the month, names of seasons, year)

- **Special days:** birthday, Easter, Christmas, Thanksgiving, vacation, holiday, school day, weekend.

ASSESSMENT

The teacher should observe the child's use of time language. She should note if he makes an attempt to place himself and events in time. Does he remember the sequence of activities at school and at home? Is he able to wait for one thing to finish before going on to the next? Is he able to order things (unit 13) in a sequence?

The following are examples of the kinds of interview tasks which are included in the Appendix.

Sample Assessment Task **Preoperational Ages 4-5**

Time (unit 15): Language Labeling and Sequence

Present the child with some pictures of daily activities such as meals, bath, nap, bedtime, etc. Ask him to tell you about the pictures: TELL ME ABOUT THIS PICTURE. WHAT'S HAPPENING? Note if he uses terms such as breakfast time, lunch time, bedtime, night, morning. After he has told you about each picture ask him, PICK OUT THE PICTURE OF WHAT HAPPENS FIRST EACH DAY. After he does that, WHAT HAPPENS NEXT? Keep asking until all the pictures are lined up. Note the child's use of time words and whether the order of the pictures makes sense.

Sample Assessment Task **Preoperational Ages 3-6**

Time and Sequence (unit 15): Identification of a Clock

Show the child a clock or a picture of a clock. WHAT IS THIS? WHAT DOES IT TELL US? Note whether he can name it and how much he can tell you about it: WHAT IS IT FOR? WHAT ARE THE PARTS AND WHAT DO THEY DO? Note if he tries to tell time.

NATURALISTIC ACTIVITIES

From birth on, time and sequence are learned. The infant learns quickly that when he wakes up from sleep, he is first held and comforted, then his diaper is changed, and then he is fed. His first sense of time duration comes from how long it takes for each of these events. The infant soon has a sense of how long he will be held and comforted, how long it takes for a diaper change, and how long it takes to eat. Time is a sense of sequence and duration of events.

The toddler shows his understanding of time words through his actions. When he is told "It's lunch time," he runs to his highchair. When he is told it is time for a nap, he may run the other way. He will notice cues which mean it is time to do something new: toys

Fig. 15-2 Dad kisses Kate goodbye when it is time to leave for work each day.

are being picked up, the table is set, or Dad appears at the door. He begins to look for these events which tell him that one piece of time ends and a new piece of time is about to start.

As spoken language develops, the child will use time words. He will make an effort to place events and himself in time. It is important for adults to listen and respond to what he has to say. The following are some examples:

- Eighteen-month-old Brad tugs at Mr. Flores' pants leg, "Cookie, cookie." "Not yet Brad. We'll have lunch first. Cookies are after lunch."

- Linda (twenty months of age) finishes her lunch and gets up. "No nap today. Play with dollies." Ms. Moore picks her up, "Nap first. You can play with the dolls later."

- "Time to put the toys away, Kate." Kate (thirty months old) answers, "Not now. I'll do it a big later on."

- Chris (three years old) sits with Mrs. Raymond. Chris says, "Last night we stayed at the beach house." "Oh yes," answers Mrs. Raymond, "You were at the beach last summer, weren't you."

(For Chris anything in the past happened 'last night').

It is very important for the young child to have a predictable and regular routine. It is through this routine that the child gains his sense of time duration and time sequence. It is also important for him to hear time words and to be listened to when he tries to use his time ideas. It is especially important that his own time words be accepted. For instance Kate's "a big later on" and Chris's "last night" should be accepted. Kate shows an understanding of the future and Chris of the past even though they are not as precise as an adult would be.

INFORMAL ACTIVITIES

The adult needs to capitalize on the child's efforts to gain a sense of time and time sequence. Reread the situations given as examples in the section before this (*Naturalistic Activities*). In each, the adults do some informal instruction. Mr. Flores reminds Brad of the coming sequence. So does Ms. Moore, so does the adult with Kate. Mrs. Raymond accepts what Chris says while at the same time she uses the correct time words "last summer." It is important that adults listen and expand on what children say.

The adult serves as a model for time related behavior. The teacher checks the clock and the calendar for times and dates. The teacher uses the time words in the *Language of Time* section. She makes statements and asks questions:

- *"Good morning*, Tom."

- *"Goodnight*, Mary. See you *tomorrow*."

- "What did you do over the *weekend*?"

- "Who is our guest for lunch *tomorrow*?"

- *"Next week* on *Tuesday* we will go to the park for a picnic."

Fig. 15-3 This father reads a book about time with his daughter.

- "Let me check the *time*. No wonder you are hungry. It's almost *noon.*"
- "You are the *first* one here *today.*"

Children will observe and imitate what the teacher says and does before they really understand the ideas completely.

STRUCTURED ACTIVITIES

Structured time and sequence activities include sequence patterns with beads, blocks and other objects, sequence stories, work centering around the calendar, and work centering around clocks. Experiences with pattern sequence and story sequence can begin at an early age. The infant enjoys looking at picture books, and the toddler can listen to short stories and begin to use beads and blocks to make his own sequences. The more structured pattern, story, calendar, clock, and other time activities described next are for children older than four and one-half.

————TIME: SEQUENCE PATTERNS————

Objective: To be able to understand and use the sequence idea of *next*

Materials: Any real things that can be easily sequenced by category, color, shape, size, etc. Some examples are listed.

- Wooden beads and strings
- Plastic eating utensils
- Poker chips or buttons or coins
- Shapes cut from cardboard
- Small toy animals or people

Activity: In this case, plastic eating utensils are used as the example. There are knives (K), forks (F) and spoons (S) in three colors (C_1, C_2, and C_3). The teacher sets up a pattern to present to the child. Many kinds of patterns can be presented. Any of the following may be used:

- Color: C_1-C_2-C_1-C_2 . . . .
  C_2-C_3-C_3-C_2-C_3-C_3 . . .
- Identity:K-F-S-K-F-S- . . . . .
  K-S-S-K-S-S . . . . . .

Say to the child: THIS PATTERN IS KNIFE, FORK, SPOON (or whatever pattern is set up). WHAT COMES NEXT? When the child has the idea of pattern then set up the pattern and say, THIS IS A PATTERN. LOOK IT OVER. WHAT COMES NEXT?

Follow-up: Do the same activity with some of the other materials suggested. Also try it with the magnet board, flannel board, and chalkboard.

Fig. 15-4 "What comes next?"

TIME: SEQUENCE STORIES

Objective: To learn sequences of events through stories

Materials: Picture story books which have clear and repetitive sequences of events, such as:

- *The Gingerbread Man*
- *The Three Little Pigs*
- *The Three Billy Goats Gruff*
- *Henny Penny*
- *Caps for Sale*

Activities: Read the stories several times until the children are familiar with them. Begin by asking, WHAT HAPPENS NEXT? before going on to the next event. Have the children say some of the repeated phrases such as: "LITTLE PIGS, LITTLE PIGS, LET ME COME IN." "NOT BY THE HAIR ON MY CHINNY-CHIN-CHIN." "THEN I'LL HUFF AND I'LL PUFF AND I'LL BLOW YOUR HOUSE IN." Have the children try to repeat the list of those who chase the Gingerbread Man. Have them recall the whole story sequence.

Follow-up: Obtain some sequence stories on cards. (such as DLM Sequential Picture Cards or Judy Sequees Puzzles)

TIME: SEQUENCE ACTIVITY, GROWING SEEDS

Objective: To experience the sequence of the planting of a seed and the growth of a plant

Materials: Radish or lima bean seeds, styrofoam cups, sharp pencil, six-inch paper plate, some rich soil, a tablespoon

Activity:

1. Give the child a styrofoam cup. Have him make a drainage hole in the bottom with the sharp pencil.
2. Set the cup on the paper plate.
3. Have the child put dirt in the cup up to about an inch from the top.
4. Have the child poke three holes in the dirt with his pointer finger.
5. Have him put one seed in each hole and cover the seeds with dirt.

6. Have the child put in one tablespoon of water.
7. Place the pots in a sunny place and watch their sequence of growth.
8. Have the children water the plants each day. Have them record how many days go by before the first plant pops through the soil.

Follow-up: Plant other types of seeds. Make a chart or obtain a chart that shows the sequence of growth of a seed. Discuss which steps take place before the plant breaks through the ground.

TIME: THE FIRST CALENDAR

Objective: To learn what a calendar is and how it can be used to keep track of time

Materials: A one week calendar is cut from posterboard with sections for each of the seven days identified by name. In each section, tabs are cut with a razor blade to hold signs made to be slipped under the tabs to indicate special times and events or the daily weather. These signs may have pictures of birthday cakes, items seen on field trips, umbrellas to show rainy days, the sun to show fair days, and so on.

Activities: Each day the calendar can be discussed. Key questions may include:

- WHAT IS THE NAME OF TODAY?
- WHAT IS THE NAME OF YESTERDAY?
- WHAT IS THE NAME OF TOMORROW?
- WHAT DAY COMES AFTER _____ ?
- WHAT DID WE DO YESTERDAY?
- DO WE GO TO SCHOOL ON SATURDAY AND SUNDAY?
- HOW MANY DAYS UNTIL _____ ?
- HOW MANY DAYS OF THE WEEK DO WE GO TO SCHOOL?
- WHAT DAY OF THE WEEK IS THE FIRST DAY OF SCHOOL?
- WHAT DAY OF THE WEEK IS THE LAST DAY OF SCHOOL?

Fig. 15-5 "What day is today?"

Follow-up: Eventually a monthly calendar can be introduced and each month's pages can be attached together and saved so that the previous month's events can be reviewed.

——TIME: THE USE OF THE CLOCK——

Objective: To find out how we use the clock to tell us when it is time to change activity

Materials: School wall clock and a handmade or a purchased large clockface such as that made by the Judy Company

Activity: Point out the wall clock to the children. Show them the clockface. Let them move the hands around. Explain how the clockface is made just like the real clockface. Show them how you can set the hands on the clockface so that they are the same as the ones on the real clock. Each day set the clockface for important times (such as clean-up, lunch, time to get up from the nap, etc.). Explain that when the real clock and the clockface have their hands in the same place that it will be time to (do whatever the next activity is).

Follow-up: Do this every day. Soon each child will begin to catch on and check the clocks. Instead of asking "When do we get up from our nap?," they will be able to check for themselves.

——TIME: BEAT THE CLOCK GAME——

Objective: To learn how time limits the amount of activity that can be done

Materials: Minute Minder or similar timer

Activities: Have the child see how much of some activity can be done in a set number of minutes:

1. How many pennies can be put in a penny bank one at a time?

2. How many times can he bounce a ball?

3. How many paper clips can he pick up one at a time with a magnet?

4. How many times can he move across the room: walking, crawling, running, going backwards, sideways, etc.? Set the timer for three to five minutes. When the bell rings the child must stop. Then count to find out how much was accomplished.

Follow-up: Try many different kinds of activities and different lengths of time. Have several children do the tasks at the same time. Who does the most in the time given?

TIME: DISCUSSION TOPICS ——FOR LANGUAGE——

Objectives: To develop time word use through discussion

Materials: Pictures collected or purchased. The following book may be used for this purpose:

Rutland, J. *Time.* New York: Grosset and Dunlap, 1976. ($1.95)

Pictures could show:

- Day and night

- Activities which take a long time and a short time

- Picture sequences which illustrate times of day, yesterday, today, and tomorrow

- Pictures which illustrate the seasons of the year
- Pictures which show early and late

Activities: Discuss the pictures using the key time words.

Follow-up: Put pictures on the bulletin board which the children can look at and talk about during their free playtime.

EVALUATION

The teacher should note whether the child's use of time words increases. She should also note whether his sense of time and sequence develops to a more mature level: Does he remember the order of events? Can he wait until one thing is finished before he starts another? Does he talk about future and past events? How does he use the calendar?

The clock? The sequence stories? The teacher may use the individual interview tasks in the Appendix.

SUMMARY

The young child can begin to learn that time has duration and that time is related to sequences of events. The child first relates time to his personal experience and to his daily sequence of activities. It is not until the child enters the concrete operations period that he can use units of time in the ways that adults use them.

The young child learns his concept of time through naturalistic and informal experiences for the most part. When he is around the age of four and one-half or five, he can do structured activities also.

SUGGESTED ACTIVITIES

- Observe preschoolers in group play. Write down any examples of time measurement activities that happened. Which stage of measurement did each represent?
- With a small group of classmates, discuss some ways a home or school environment can encourage children to use time measurement skills.
- Plan some structured time sequence and time duration measurement activities. Try to use one with a small group of preschoolers. Share with the class what happened.
- Add structured time sequence and duration measurement activities to the Math Resource File.

REVIEW

A. Describe what is meant by the sequence of time.

B. Describe what is meant by the duration of time.

C. Listed below are comments made by children. Indicate whether the statement represents (a) sequence of time, (b) duration of time, or (c) none of these.

1. Child with ball says, "It went up."
2. Child says, "Mama read me a story at bedtime."
3. Child says, "I stayed at grandma's for two nights."
4. Child says, "Daddy plays with me for hours and hours."
5. Child says, "It's snack time."
6. Child says, "This ball weighs two tons."
7. Child says, "Teacher, this picture took me a long time to paint."

D. List the three kinds of time and give examples of each.

E. Match each item in Column II with the correct item in Column I.

I	II
1. General time word	a. Tomorrow
2. Specific time word	b. Two hours
3. Relational word	c. This morning
4. Specific duration word	d. Christmas
5. Special days	e. Once
	f. Three years old
	g. Birthday
	h. One minute
	i. Yesterday

unit 16 graphs

OBJECTIVES

After studying this unit, the student should be able to

- Explain the use of graphs
- Describe the three stages that young children go through in making graphs
- List materials to use for making graphs.

Ms. Moore hears George and Sam talking in loud voices. She goes near them and hears the following discussion.

George: "More kids like red than blue."

Sam: "No! No! More like blue!"

George: "You are all wrong."

Sam: "I am not. You are all wrong."

Ms. Moore goes over to the boys and asks, "What's the trouble, boys?" George replies, "We have to get paint to paint the house Mr. Brown helped us build. I say it should be red. Sam says it should be blue."

Sam insists, "More kids like blue than red."

Ms. Moore suggests, "Maybe there is some way we can find out." She takes the boys to a table.

"On this table let's put a piece of blue paper and a piece of red paper and a bowl of red and blue cube blocks."

George's eyes light up, "I see, and then each child can vote, right?"

Sam and George go around the room. They explain the problem to each child. Each child comes over to the table. They each choose one block of the color they like better and stack it on the paper of the same color. When they finish, there are two stacks of blocks as shown in figure 16-1.

Ms. Moore asks the boys what the vote shows. Sam says, "The red stack is higher. More children like the idea of painting the house red." "Good," answers Ms. Moore, "would you like me to write that down for you?" Sam and George chorus, "Yes!"

"I have an idea," says George, "Let's

make a picture of this for the bulletin board so everyone will know. Will you help us Ms. Moore?"

Ms. Moore shows them how to cut out squares of red and blue paper to match each of the blocks used. The boys write red and blue on a piece of white paper and then paste the red squares next to the word red and the blue squares next to the word blue. Ms. Moore shows them how to write the title: Choose the Color for the Playhouse. Then they glue the description of the results at the bottom. The results can be seen in figure 16-2.

In the preceding example, the teacher helped the children solve their problem by making two kinds of graphs. Graphs are used to show visually two or more comparisons in a clear way. When a child makes a graph, he uses basic skills of classifying, comparing, counting, and measuring to make a picture of some information. A child who has learned the basics of math will find this to be an interesting and challenging activity.

STAGES OF DEVELOPMENT FOR MAKING AND UNDERSTANDING GRAPHS

The type of graphs that a child can make and understand change as the child grows. The child goes through five stages. The first three stages are described in this unit. The fourth and fifth are included in unit 20. In stage one, the child uses real objects to make his graph. Sam and George used inch-cube

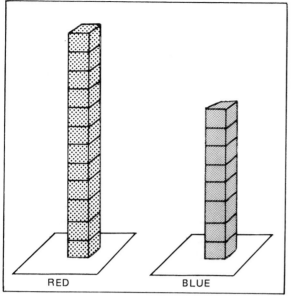

Fig. 16-1 A three-dimensional graph which compares children's preferences for red or blue

blocks. At this stage only two things are compared. The main basis for comparison is one-to-one correspondence (one block for each child).

In the second stage, more than two items are compared. In addition, a more permanent record is made — such as when Sam and George in an earlier example glued squares of paper on a piece of paper for the bulletin board. An example of this type of graph is shown in figure 16-3, page 125. The teacher has lined off twelve columns on poster board (or large construction paper). Each column stands for one month of the year. Each child is given a paper circle. Crayons, water markers, glue, and yarn scraps are available so each child can draw his own head and place it on the month for his birthday. When each child

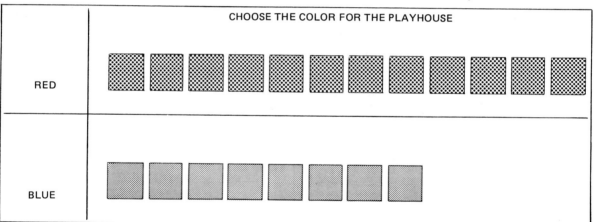

The red row is longer than the blue row. More children like red than like blue. We will buy red paint for our playhouse. 12 like red. 8 like blue.

by George and Sam

Fig. 16-2 The color preference graph is copied using squares of red and blue paper.

Jan.	Feb.	March	April	May	June	July	August	Sept.	Oct.	Nov.	Dec.

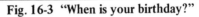

April has the most birthdays. There are four.
March and October have no birthdays.
Three months have three.
One month has two.
Five months have one.

Fig. 16-3 "When is your birthday?"

has put his head on the graph, the children can compare the months to see which month has the most birthdays.

In the third stage, the children progress through the use of more pictures to block charts. They no longer need to use real objects but can start right off with cut out squares of paper. Figure 16-4 shows this type of graph. In this stage, the children work more independently.

DISCUSSION OF A GRAPH

As the children talk about their graphs and dictate descriptions for them, they use math words. They use words such as:

less than	the same as
more than	none
fewer than	all
longer, longest	some
shorter, shortest	a lot of
the most	higher
the least	taller

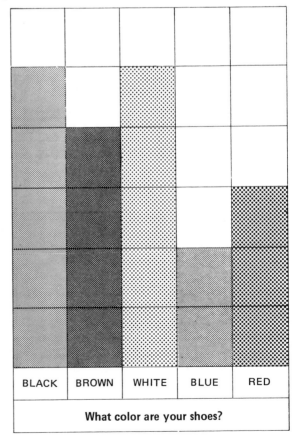

Fig. 16-4 A Block Graph Made with Paper Squares.

125

MATERIALS FOR MAKING GRAPHS

There are many kinds of materials that can be used for the first stage graphs. An example has been shown in which cube blocks were used. Other materials can be used just as well.

At first it is best to use materials that can be kept in position without being knocked down or pushed apart by young children. Stands can be made from dowel rods. A washer or curtain ring is then placed on the dowel to represent each thing or person, figure 16-5(A). Strings and beads can be used. The strings can be hung from hooks or a rod; the lengths are then compared, figure 16-5(B). Unifix interlocking cubes, figure 16-5(C), or pop beads, figure 16-5(D), can also be used.

Once the children have worked with the more stable materials, they can use the cube blocks and any other things which can be lined

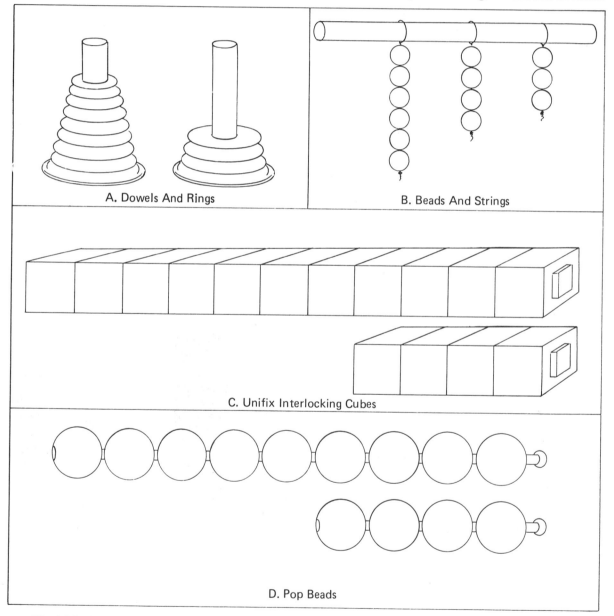

A. Dowels And Rings

B. Beads And Strings

C. Unifix Interlocking Cubes

D. Pop Beads

Fig. 16-5 Four Examples of Three-Dimensional Graph Materials which Can Be Made into a Stable Graph.

up. Poker chips, bottle caps, coins, spools, corks, and beans are good for this type of graph work, figure 16-6.

At the second stage, graphs can be made with these same materials but with more comparisons made. Then the child can go on to more permanent recording by gluing down cut out pictures or markers of some kind.

At the third stage, the children can use paper squares. This prepares the way for the use of squared paper. (This will be included in unit 20.)

TOPICS FOR GRAPHS

Once children start making graphs, they often think of problems to solve on their own. The following are some comparisons that might be of interest:

- number of brothers and sisters
- hair color, eye color, clothing colors
- kinds of pets children have
- heights of children in the class
- number of children in class each day
- sizes of shoes
- favorite TV programs (or characters)
- favorite foods
- favorite colors
- favorite storybooks
- type of weather each day for a month
- number of cups of water or sand which will fill different containers
- time in seconds, to run across the playground

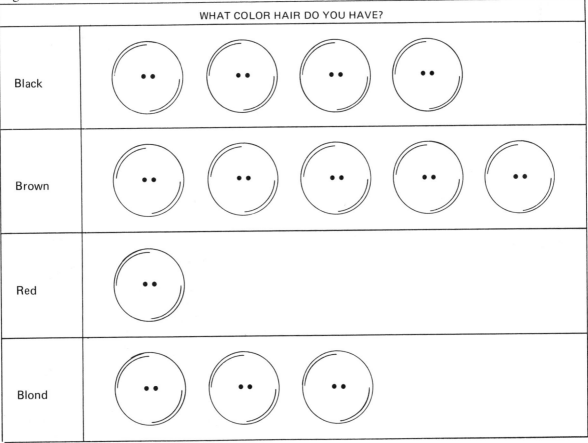

Fig. 16-6 Graph Made with Buttons Glued to Cardboard

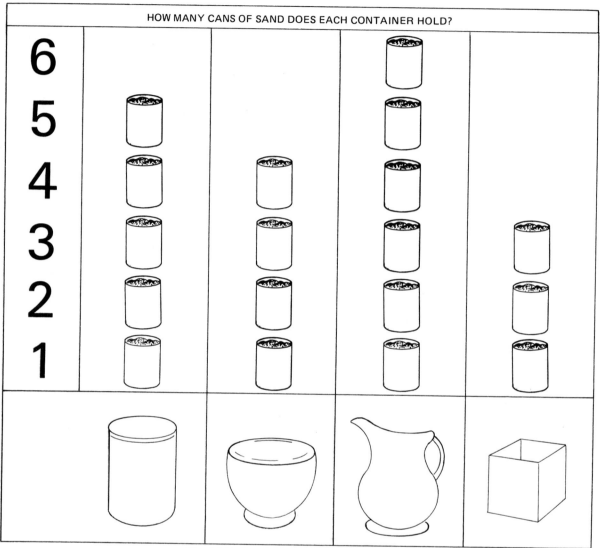

HOW MANY CANS OF SAND DOES EACH CONTAINER HOLD?

Fig. 16-7 Graph Made with Paper Cutouts

SUMMARY

Making graphs provides a means for use of some of the basic math skills in a creative way. Children can put into a picture form the results of classifying, comparing, counting, and measuring activities.

The first graphs are three dimensional made with real objects. The next are made with pictures and the next with paper squares. Children can discuss the results of their graph projects and dictate a description of the meaning of the graph to be put on the bulletin board with the graph.

Fig. 16-8 These girls use inch cubes to make a three-dimensional graph.

SUGGESTED ACTIVITIES

- Discuss with some preschool teachers how they use graphs in their schools. Share with the class what was learned.
- Collect materials from which to make graphs.
- If possible, supply a group of children with graph materials and topics, and observe what kinds of graphs they make.
- Add graph activities to the Activities File.

REVIEW

A. Explain the use of graphs.

B. List and describe the three stages of development for making and understanding graphs.

C. Place the following types of graphs in the order in which young children are able to make and understand them.

   a. picture graphs
   b. paper square graphs
   c. real object graphs

D. List five materials to use for making graphs.

unit 17 practical activities

OBJECTIVES

After studying this unit, the student should be able to

- Describe how children learn math through dramatic role playing
- Describe how children learn math through food experiences
- Encourage dramatic role playing that relates to helping children learn math

Math skills and ideas are only valuable to the child if he can use them in his everyday life. The young child spends most of his waking time involved in play. Play can be used as a vehicle for the use of math ideas. The young child also likes to feel he is big and can do "big person" things. He likes to pretend he is grown up, and he wants to do as many of the same things grownups do as he can. Role playing can be used as means for the child to apply what he knows about math. Experiences with food also include a variety of opportunities to do things that big people do. The child can plan meals, snacks, and parties. He can grow, buy, cook, serve, and eat foods. For the teacher, these activities are a time when she can assess and evaluate through observation.

DRAMATIC ROLE PLAYING

When the child does dramatic role playing, he practices what it is like to be big. He begins with simple imitation of what he has seen. His first roles reflect what he has seen at home. He bathes, feeds, and rocks babies. He cooks meals, sets the table, and eats. Usually one of his first outside experiences is go to the store to shop. This is soon reflected in the roles he plays. He begins with carrying things in bags and purses. At first anything will do that will fit in his bag or purse. Soon he wants more realistic props such as play

money and empty food containers. Next he may build a store from big blocks and boards. He will soon learn to play cooperatively with other children. One will be the mother, another the father, another the child, and another a store clerk. When the child reaches this stage, the teacher can expand the store play by providing more props and background experiences.

Background experiences include trips, discussions, stories, records, and films. Visits

Fig. 17-1 Going shopping is an important job.

130

can be made to stores, to restaurants, to the post office, to the bank and other service centers. Stories can be read, trips discussed, records played, and films shown which tell more and add to what the child has experienced. Basic props for shopping include play money, a cash register, purses, wallets, and bags and baskets for carrying the things which are bought. Each type of store or business can then be set up with props which fit the items or services to be bought.

Some stores and other places which inspire the use of math skills and ideas are listed below.

- Toy Store: Children can bring old toys from home or use toys already in the school to buy and sell.

- Grocery Store: Children and teachers can bring empty food containers from home to stock the store. They can make pretend food from play dough and papier mache. Plastic food can also be used.

- Clothing Store: Old clothes can be brought. Departments can be set up for baby clothes, children's clothes, ladies' clothes, and men's clothes. A shoe section can be included.

- Jewelry Store: Old jewelry can be brought from home. Jewelry can be made by stringing paper or macaroni. Watches and clocks can be made from paper.

- Service Centers: Post office, bank, gasoline station, auto repair shop, beauty shop, barbershop, doctor's office, restaurant, and fast food business.

- Transportation: Bus, plane, train, and boat trips. Children can buy tickets.

Math skills and ideas are used in all these play activities. Some examples would include the following:

- Matching: Exchanging money for goods or services. The child gives one or more pieces of play money for a gallon of gasoline, box of cereal, haircut, postage stamp, etc. which he buys.

- Counting: Things which are bought are counted. He buys one coat, two bottles of milk, three oranges, ten gallons of gasoline, three hamburgers, and so on. The money paid must also be counted.

- Comparing and measuring: Clothing can be tried on for fit. Real tape measures and rulers can be used to "measure" feet, shoulders, arms, and other body parts for fit. Meat, fruits, and vegetables can be weighed. The doctor and nurse can check the sick person's temperature. Things can be available in containers of different sizes so that large, small, or middle sized can be bought. Milk cartons can be labeled as quarts, half gallons, and gallons. Customers may ask for a little bit or a lot.

- Spatial relations and volume: Things bought must be fit into bags, boxes, and baskets that will hold them.

- Sets and Classifying: Things brought home must be put in the right place in the right room. To set up a store requires sorting. For example, the grocery store must have places for meat, for dairy products, for frozen food, for fresh food, for cans, for boxes, for soaps, and so on.

- Number Symbols: For the child who can name the number symbols, price tags can be made. The child who can write number symbols can make price tags and write bills of sale. They can be shown how to make dollar ($) and cents (¢) symbols for their tags. They can write checks and make bank deposits. Number symbols can also be put on play coins and bills.

FOOD EXPERIENCES

Food offers many activities which use math. Foods are familiar things to the child. From infancy he is, of course, learning about food. He has many naturalistic experiences buying, preparing, serving, and eating food. He may also have grown food or seen it grown if he lives where there is a garden or in a rural area where there are farms.

In the housekeeping area at school, the child cooks, serves, and eats food in his role playing. It was suggested in unit 14 that a simple measuring activity could be to make flour and salt play dough. This dough can then be made into pretend cookies, cakes, and other foods to be used as props for dramatic role playing.

The young child can also make real things to eat. He can learn that certain amounts of each item to be used in the recipe must be carefully measured in order to come out with something that looks and tastes good to eat. This is an excellent way to introduce the child to the concept of standard units in a natural and informal way. As young as two or three, the child can understand that the recipe says that only so much is to be used. He can be shown that we have special cups and spoons which are used to measure food. In this country, we measure food by volume (cups, teaspoons, and tablespoons). In some other countries, food is measured out by weight. For instance, compare the following recipes, one from England and one from the United States.

From England: Fairy Cakes[1]
8 oz flour
4 oz fat
4 oz sugar
1 or 2 eggs

Fig. 17-2 **The children measure when they make sugar cookies.**

From the U.S.A.: Play Dough Biscuits[2]
2 cups sifted unbleached white flour
3 3/4 teaspoons baking powder
1 teaspoon salt
1/3 cup oil
3/4 cup milk

Figures 17-3, 17-4, and 17-5 illustrate three popular recipes that young children enjoy making. In addition to measuring, the children count (number of cups, teaspoons, ounces). They also learn to follow a sequence of steps such as measure, mix, bake, cut, and serve. For things that are baked, they learn that a certain temperature must be used and that the baking takes a special length of time. They learn that some foods need a high heat and some a low heat and that some must be chilled and others frozen. Some things are cut into two or more pieces when they are prepared. The finished food must be divided into portions to be served.

The child who lives where food grows has even more opportunities for the use of math skills. The eggs laid by the hens can be

[1]*Beginnings* (New York: John Wiley and Sons Inc., 1968).

[2]Doreen Croft and Robert Hess, *An Activities Handbook for Teachers of Young Children,* Second edition (Boston: Houghton Mifflin Company, 1975).

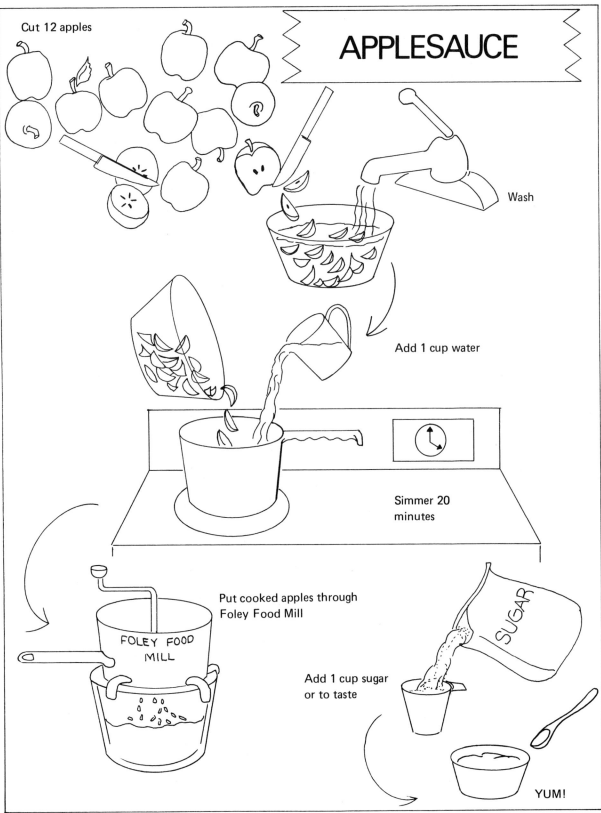

Cut 12 apples

APPLESAUCE

Wash

Add 1 cup water

Simmer 20 minutes

Put cooked apples through Foley Food Mill

FOLEY FOOD MILL

Add 1 cup sugar or to taste

SUGAR

YUM!

Fig. 17-3 Recipe for Applesauce

counted each day. A graph can be made showing the comparison of the number of eggs collected each day. The eggs can be sold, and the children can record the amount of money received. The number of days from the time each vegetable in the garden is planted until it is picked can be recorded. The number of carrots, beans, squash, and so on that are picked each day can be recorded also.

When the child sets the table in the play area or for a real meal, he uses math skills.

He counts and matches the number of people to be served, the number of spaces at the table, the napkins, dishes, spoons, forks, knives, and cups. As he cleans up and puts things away, he sorts things into sets. A silverware holder with a place for each eating utensil offers a structured sorting experience. The shelves for each type of dish and bowl and cup can be labeled with pictures to help the children match and sort them into the right places.

FIRST DAY

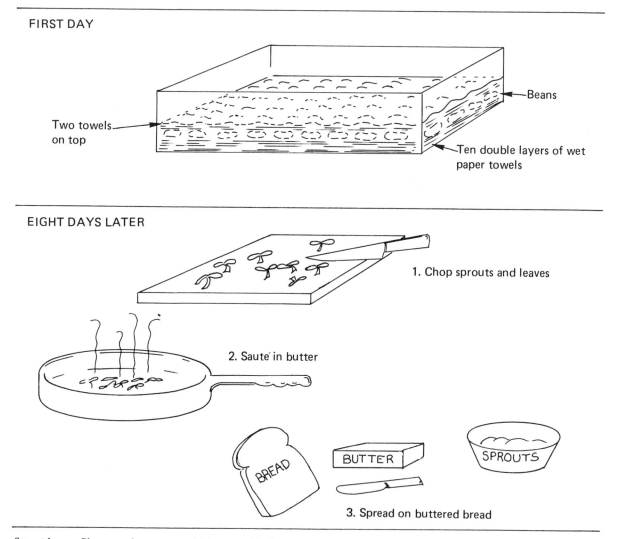

EIGHT DAYS LATER

Sprout beans. Place navy beans on ten thicknesses of 2-ply paper towels, covered by two thicknesses of paper towels. Keep beans moist and warm for about eight days or until leaves appear on 1 1/2 inch sprouts. Chop beans, sprouts, and leaves until fine. Saute in butter and salt. Spread on buttered bread.

Fig. 17-4 Bean Sprout Sandwich

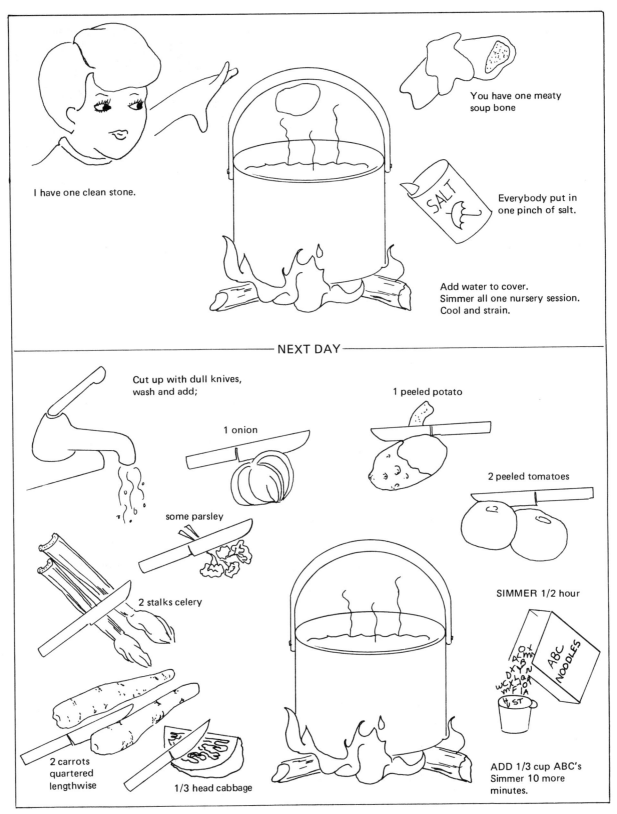

I have one clean stone.

You have one meaty soup bone

Everybody put in one pinch of salt.

Add water to cover.
Simmer all one nursery session.
Cool and strain.

— NEXT DAY —

Cut up with dull knives, wash and add;

1 onion

1 peeled potato

2 peeled tomatoes

some parsley

2 stalks celery

SIMMER 1/2 hour

2 carrots quartered lengthwise

1/3 head cabbage

ADD 1/3 cup ABC's
Simmer 10 more minutes.

Fig. 17-5 Recipe for Stone Soup

SUMMARY

Dramatic role play and food experiences give the child practical uses for his math skills. As he plays house, store, and service roles, the child can match, count, classify, compare, measure, and use spatial relations concepts and number symbols. He also practices the exchange of money for goods and services.

Through food experiences the child can learn the sequence from planting and growing to picking, buying, cooking, serving, and eating. He can also count, measure, and match to serve a real purpose.

For the teacher, these activities offer valuable times to observe the child and assess his ability to use math in everyday situations.

Fig. 17-6 Scott spreads peanut butter on one piece of bread and jelly on the other. Two pieces put together make a sandwich.

SUGGESTED ACTIVITIES

- Observe children ages three, four, and five at play in school. Note if any math experiences take place during dramatic role playing or while working with foods. Share what you observed with your class.
- Add five dramatic role playing math activities to the Activities File.
- Review cookbooks and examine recipes recommended for young children.
- Add five food experiences that use math to the Activities File.
- If possible, view the film "Jenny Is A Good Thing." Information about obtaining the film on a free loan basis is available from Sponsor Service Desk, Modern Talking Picture Service, Inc., 1212 Avenue of the Americas, New York, New York 10036.

REVIEW

A. Briefly answer each of the following.

1. Why should dramatic role playing be included as a math experience for young children?

2. Why should food activities be included as math for young children?

3. How can the teacher encourage dramatic role playing that includes math experiences?

4. What are four food activities that can be used in the preschool math program?

B.   Indicate which item (a) or (b) in Column II matches each activity in Column I.

I

1. Planting bean seeds
2. Buying bananas
3. Using a cash register
4. Making and selling lemonade
5. Having an auto repair shop
6. Setting the snack table

II

a. Dramatic Role Play Experience
b. Food Experience

Section 4 Symbols and Higher Level Activities

unit 18 symbols

OBJECTIVES

After studying this unit, the student should be able to

- List the six number symbol skills
- Describe four basic types of self-correcting math materials
- Do structured number symbol activities with children

Number symbols are called numerals. Each numeral represents an amount and acts as a shorthand for recording *how many*. The young child sees numerals all around. He has some idea of what they are before he can understand and use them. He sees that there are numerals on his house, the phone, the clock, and the car license plate. He may have

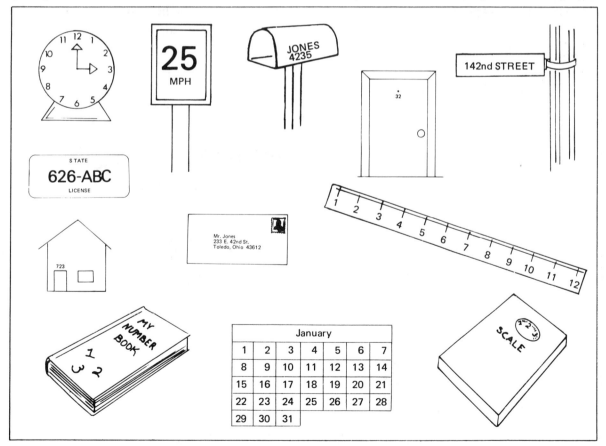

Fig. 18-1 Numerals are seen everywhere in the environment.

one or more counting books. He may watch a children's TV program where numeral recognition is taught. Sometime between the age of two and the age of five a child learns to name the numerals from zero to ten. However, the child is usually four or more when he begins to understand that each numeral stands for a set of things of a certain amount that is always the same. He may be able to tell the name of the symbol "3" (three) and count three objects, but he may not realize that the "3" can stand for the three objects. This is illustrated in figure 18-2.

It can be confusing to the child to spend time on drill with numerals until he has had many concrete experiences with basic math concepts. Most experiences with numerals should be naturalistic and informal.

THE NUMBER SYMBOL SKILLS

There are six number symbol skills that the young child learns:

- He learns to recognize and say the name of each numeral.

- He learns to place the numerals in order: 0-1-2-3-4-5-6-7-8-9-10.

- He learns to associate numerals with sets: "1" goes with one thing.

- He learns that each numeral in order stands for one more than the numeral that comes before it. (That is, two is one more than one, three is one more than two, and so on.)

- He learns to match each numeral to any set of the size that the numeral stands for and to make sets that match numerals.

- He learns to reproduce (write) numerals.

The first four skills are included in this unit. The last two are the topics for unit 19.

ASSESSMENT

The teacher should observe whether the child shows an interest in numerals. Does he repeat the names he hears on television? Does he use self-correcting materials that are described in the part of this unit on informal learning and teaching? What does he do when he uses these materials? Individual interviews would include the types of tasks which follow.

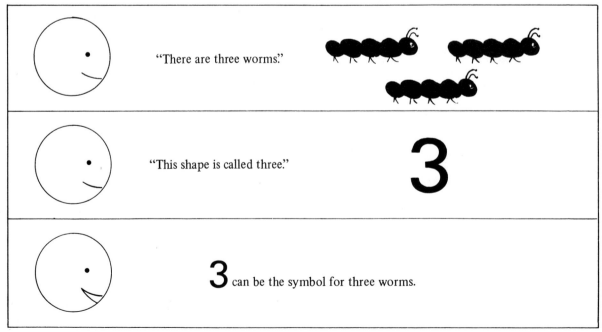

Fig. 18-2 The child counts the objects, learns the symbol, and realizes that the symbol can stand for the set.

Sample Assessment Task   Preoperational Ages 3-6

Symbols (unit 18): Recognition

Starting with zero show the child, one at a time, cards with the numerals from zero to ten. **WHAT IS THIS? TELL ME THE NAME OF THIS.**

Sample Assessment Task   Preoperational Ages 4-6

Symbols (unit 18): Sequencing

Have eleven cards. On each card is one of the number symbols from zero to ten. Place them all in front of the child in random order. Say: **PUT THESE IN ORDER. WHICH COMES FIRST? . . . . . NEXT? . . . NEXT?**

Sample Assessment   Preoperational Ages 5 and older Task

Symbols (unit 18): One More Than

Have eleven cards. On each card is one of the number symbols from zero to ten. Place them in front of the child in order from zero to ten. Ask: **WHICH NUMERAL MEANS ONE MORE THAN TWO? WHICH NUMERAL MEANS ONE MORE THAN SEVEN? WHICH NUMERAL MEANS ONE MORE THAN FOUR?** (If the child answers these, then try **LESS THAN**)

NATURALISTIC ACTIVITIES

As the young child observes his environment, he sees numerals around him. He sees them on clocks, phones, houses, books, food containers, television programs, money, calendars, thermometers, rulers, measuring cups, license plates, and on many other objects in many places. He hears people say:

- My phone number is 622-7732.
- My house number is 1423.
- My age is six.
- I have a five-dollar bill.

Usually the child starts using the names before he places them with the number symbols.

Fig. 18-3 "It's a one!"

- Diana is playing house. She takes the toy phone and begins dialing, "One-six-two. Hello dear, will you please stop at the store and buy a loaf of bread."
- "How old are you, Pete?" "I'm six," answers two-year-old Pete.
- "One, two, three, I have three dolls."
- Tanya is playing house. She looks up at the clock. "Eight o'clock and time for bed," she tells her doll. (The clock really says 9:30 A.M.).

The child begins to learn number symbols as he looks and listens and then imitates in his play what he has seen and heard.

INFORMAL ACTIVITIES

In school most activities with numerals should be informal. Experimentation and practice in perception with sight and touch are most important. These experiences are made available by means of activities with self-correcting manipulative materials. Self-correcting materials are those which the child can use by trial and error to solve a problem as he works by himself. The material is made in such a way that it can be used with success with very little help. Manipulative materials

Fig. 18-4 Putting in and taking out tactile numerals requires the use of touch and sight.

Fig. 18-5 Self-correcting materials can be used by one child.

are things which have parts and pieces which can be picked up and moved by the child to solve the problem presented by the materials. The teacher observes the child as he works. She notes whether the child works in an organized way and whether he sticks with the material until he has the task finished.

There are four basic types of self-correcting manipulative math materials that can be used for informal activities. These materials can be bought or they can be made. The four basic groups of materials are those which teach discrimination and matching, those which teach sequence (or order), those which give practice in association of symbols with sets, and those which combine association of symbols and sets with sequence. Examples of each type are illustrated in figures 18-7 through 18-10.

The child can learn to discriminate one numeral from the other by sorting packs of numeral cards. He can also learn which numerals are the same as he matches. Another type of material which serves this purpose is a lotto-type game. The child has a large card divided equally into four or more parts. He must match individual numeral cards to each

numeral on the big card. These materials are shown in figure 18-7 (A and B). He can also experiment with felt, plastic, magnet, wooden, rubber, and cardboard numerals.

There are many materials which teach sequence or order. These may be set up so that parts can only be put together in such a

Fig. 18-6 Self-correcting materials can be used cooperatively by more than one child.

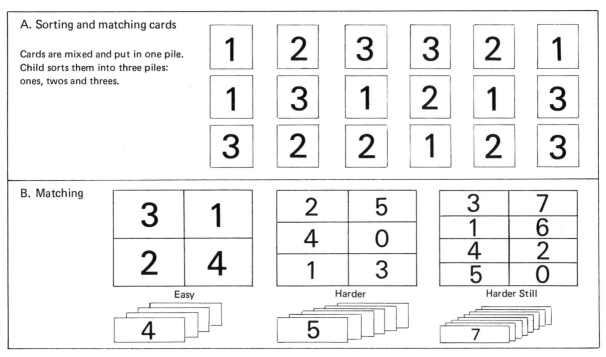

Fig. 18-7 Sorting and Matching

way that when the child is done, he sees the numerals are in order in front of him. An example would be the Number Worm® by Childcraft, figure 18-8(A). Sequence is also taught through the use of a number line or number stepping stones. The Childcraft giant Walk-On Number Line® lets the child walk from one numeral to the next in order, figure 18-8(B). The teacher can set out numerals on the floor (such as Stepping Stones® from Childcraft) which the child must step on in order, figure 18-8(C).

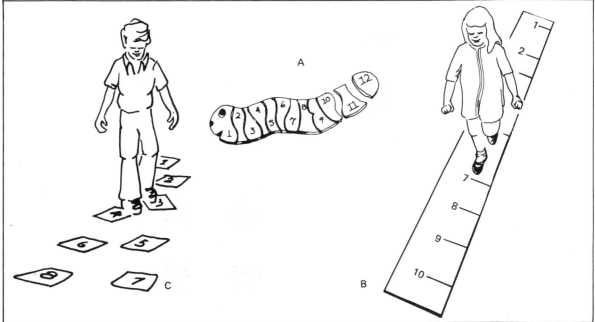

Fig. 18-8 Materials That Help the Child Learn Numeral Sequence

Many materials can be bought which help the child associate each numeral with the set that goes with it. Large cards which can be placed on the bulletin board (such as Childcraft Poster Cards®) give a visual association, figure 18-9(A). Numerals can be seen and touched on textured cards which can be bought (such as Childcraft Beaded Number Cards®). Numeral cards can be made using sandpaper for the sets of dots and for the numerals, figure 18-9(B). Other materials require the child to use visual and motor coordination. He may have to match puzzle-like pieces (such as Math Plaques® from Childcraft), figure 18-9(C). He may put pegs in holes (such as Peg Numerals® from Childcraft), figure 18-9(D). Unifix® inset pattern boards require the same type of activity. The teacher can make cards which have numerals and dots the size of buttons or other counters. The child could place a counter on each dot, figure 18-9(E).

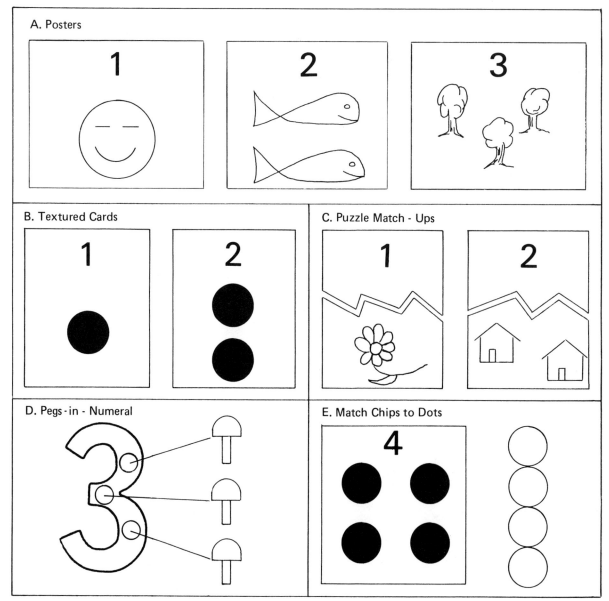

Fig. 18-9 Numeral and Set Association

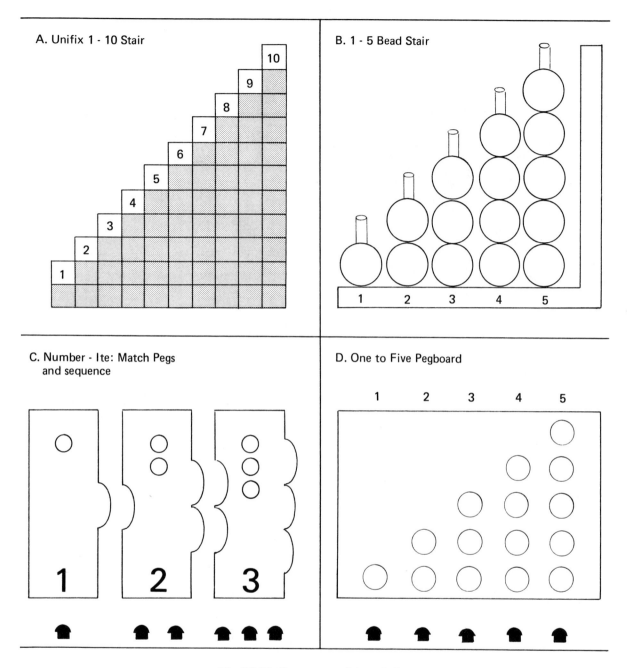

A. Unifix 1 - 10 Stair

B. 1 - 5 Bead Stair

C. Number - Ite: Match Pegs and sequence

D. One to Five Pegboard

Fig. 18-10 Sequence and Association

Materials which give the child experience with sequence and association at the same time are also available. These are shown in figure 18-10. It can be seen that the basis of the materials is that the numerals are in a fixed order and the child adds some sort of counter which can only be placed in the right amount. Unifix stairs are like the inset patterns but are stuck together, figure 18-10(A). Other materials illustrated are counters on rods, figure 18-10(B), pegs in holes (Number-Ite® from Childcraft), figure 18-10(C), or 1-5 pegboard, figure 18-10(D).

The teacher's role with these materials is to show the child how they can be used and then step back and watch. After the child has

Fig. 18-11 Cube blocks can be used informally to match numerals.

learned to use the materials independently, the teacher can make comments and ask questions:

- How many pegs on this one?

- Can you tell me the name of each numeral?

- You put in the four pegs that go with that numeral four.

- How many beads are there here? (point to stack of one) How many here? (stack of two) (and so on.)

- Good, you separated all the numerals into piles. Here are all "ones" and here are all "twos."

Most children will learn through this informal use of materials to recognize and say the name of each numeral, to place the numerals in order, to see that each numeral stands for one more than the one before it, to associate numerals with amounts. However, some children will need the structured activities described next.

STRUCTURED ACTIVITIES

By the time the young child enters kindergarten, he should be able to

- Recognize the numerals from zero to ten

- Place the numerals from zero to ten in order

- Know that each numeral represents a set one larger than the numeral before (and one less than the one that comes next)

- Know that each numeral represents a set of things

Numeral	Amount in Set
0	
1	X
2	XX
3	XXX
4	XXXX
5	XXXXX
6	XXXXXX
7	XXXXXXX
8	XXXXXXXX
9	XXXXXXXXX
10	XXXXXXXXXX

He may not always match the right numeral to the correct amount, but he will know that there is such a relationship. The five-year-old child who cannot do one or more of the tasks listed needs some structured help.

———NUMERALS: RECOGNITION———

Objective: To learn the names of the number symbols

Materials: Write the numerals from zero to ten on flash cards.

Activity: This is an activity that a child who can name all the numbers can do with a child who needs help. Show the numerals one at a time in order, THIS NUMERAL IS CALLED _____ . LET'S SAY IT TOGETHER, ___ . Do this for each numeral. After ten, say, I'LL HOLD THE CARDS UP ONE AT A TIME.

Fig. 18-12 The flannel board numerals can be moved until they are in the correct sequence.

YOU NAME THE NUMERAL. Go through once. Five minutes at a time should be enough.

Follow-up: Give the child a set of flash cards to review on his own.

NUMERALS: SEQUENCE AND ONE MORE THAN

Objective: To learn the sequence of numerals from zero to ten

Materials: Flannel or magnet board, felt or magnet numerals, felt or magnet shapes (such as felt primary cutouts, or magnet geometric shapes)

Activity: Put the "zero" up first at the upper left-hand corner of the board. WHAT IS THIS NUMERAL CALLED? If the child cannot tell you, say: THIS IS CALLED "ZERO." Put the "one" numeral up next to the right. WHAT IS THIS NU-MERAL CALLED? If the child cannot tell you, say: THIS IS "ONE." SAY IT WITH ME, "ONE." Continue to go across until the child does not know two numerals in a row. Then, go back to the beginning of the row. TELL ME THE NAME OF THIS NUMERAL. YES, "ZERO." WHAT IS THE NAME OF THE NEXT ONE? YES, IT IS "ONE" SO I WILL PUT ONE RABBIT HERE. Put one rabbit under the "one." THE NEXT NUMERAL IS ONE MORE THAN "ONE." WHAT IS IT CALLED? After

the child says "two" on his own or with your help, let him pick out two shapes to put on the board under the "two." Keep going across until you have done the same with each numeral he knows, plus two that he does not know.

Follow-up: Have the child set up the sequence. If he has trouble, ask: WHAT COMES NEXT? WHAT IS ONE MORE THAN _____ ? Leave the board and the numerals and shapes out during playtime. Encourage the children who know how to do this activity to work with a child who does not.

NUMERALS: RECOGNITION, SEQUENCE, ASSOCIATION WITH SETS, ONE MORE THAN

Objective: To help the child to integrate the concepts of association with sets, with one more than, while learning the numeral names and sequence

Materials: Cards with numerals zero to ten and cards with numerals and sets zero to ten

Activities:

1. I'M GOING TO PUT DOWN CARDS WITH EACH NUMERAL ON IT — UP TO TEN. SAY THE NAMES WITH ME IF YOU KNOW THEM.

2. HERE IS ANOTHER SET OF CARDS WITH NUMERALS. Give the cards with numerals and sets to the child. MATCH THESE UP WITH THE OTHER CARDS. LET'S SAY THE NAMES AS YOU MATCH.

Follow-up: Let the child do this activity on his own. Encourage him to use the self-correcting materials also.

EVALUATION

The teacher may question each child as he works with the self-correcting materials. She should note which numerals he can name and if he names them in sequence. She may interview him individually using the assessment questions in the Appendix.

SUMMARY

Numerals are the symbols used to represent amounts. The name of each numeral must be learned. The sequence, or order, must also be learned. The child needs to understand that each numeral represents a set that is one larger than the one before (and one less than the one that comes next).

Most children learn the properties and purposes of numerals through naturalistic and informal experiences. There are many excellent self-correcting materials which can be bought or made for informal activities. Any structured activities should be short in length.

SUGGESTED ACTIVITIES

- Observe children to see how they use numerals. Share your experience with your class.

- Talk with preschool and kindergarten teachers about how they teach structured number symbol skills. Share with your class what you learned.

- Add number symbol skill activities to your Resource File.

- Make a number symbol skill activity material and share it with the class.

- Go to a toy store and examine the materials available that are used to teach number symbols.

- Do a structured number symbol activity, first with one child and, then, with a small group of children.

REVIEW

A. Answer each of the following.

1. What are numerals?

2. What are the six number symbol skills?

3. What is meant by self-correcting manipulative materials?

4. What are the four basic types of self-correcting manipulative materials used to teach number symbol skills?

5. What is the teacher's role when using self-correcting manipulative materials?

B. Below are listed some numeral materials. Tell what they could help a child learn in the numbers skill area.

1. Magnet board, 0-5 numerals, various magnetic animals
2. Matchmates® (match plaques)
3. A set of eleven cards containing the number symbols 0 to 10
4. Lotto® numeral game
5. Number Worm®
6. Pegs-in-Numerals®
7. Unifix 1-10 Stair®
8. Walk on Line®
9. Number-Ite®
10. 1-2-3 Puzzle®

unit 19 sets and symbols

OBJECTIVES

After studying this unit, the student should be able to

- Describe the three higher level learnings of sets and symbols
- Plan and do structured sets and symbols activities with young children
- Assess and evaluate a child's ability to use math sets and symbols

The activities in this unit build on many of the ideas and skills presented in earlier units: matching, number and counting, sets and classifying, comparing, ordering, and symbols.

The experiences in this unit will be most meaningful to the child who can

- Match things one-to-one and match sets of things one-to-one

- Recognize sets of one to four without counting and count sets up to at least ten things without a mistake

- Divide large sets into smaller sets and compare sets of different amounts

- Place sets containing different amounts in order from least to most

- Name each of the numerals from zero to ten

- Recognize each of the numerals from zero to ten

- Be able to place each of the numerals in order from zero to ten

- Understand that each numeral stands for a certain number of things

- Understand that each numeral stands for a set of things one more than the numeral before it and one less than the numeral after it

When the child has reached the objectives in the preceding list, he can then learn

- To match a symbol to a set. That is, if he is given a set of four items, he can pick out or write the numeral *4* as the one that goes with that set.

- To match a set to a symbol. That is, if he is given the numeral *4*, he can make or pick out a set of four things to go with it.

- To reproduce symbols. That is, he can learn to write the numerals.

Fig. 19-1 As mother watches, Lisa puts the car with two dots on the hood in the garage with the numeral "2" on the roof.;

ASSESSMENT

If the child can do the assessment tasks in units 5-8, 13, and 18, then he has the basic skills to learn the skills and ideas included in this unit. In fact, he could probably be observed doing some symbol and set activities on his own if materials are made available during free playtime. Individual interviews would include the following types of tasks.

Sample Assessment Preoperational/Concrete Ages 5-7 Task

Sets and Symbols (unit 19): Matching a Symbol to a Set

Lay out in front of the child cards on which are written the numerals from zero to ten. One at a time, show the child sets of each amount. These sets can be objects or can be drawn on cards or both. **PICK OUT THE NUMERAL THAT TELLS HOW MANY THINGS ARE IN THIS SET.**

Sample Assessment Preoperational/Concrete Ages 5-7 Task

Sets and Symbols (unit 19): Matching a Set to a Symbol

Lay out in front of the child cards on which are written the numerals from zero to ten. Have a container of counters (such as chips, buttons, inch-cube blocks). **MAKE A SET FOR EACH NUMERAL.**

Sample Assessment Preoperational/Concrete Ages 5-7 Task

Sets and Symbols (unit 19): Reproduce (Write Numerals

Give the child a pencil, crayon, or marker and a piece of paper. **WRITE AS MANY NUMBERS AS YOU CAN.** Note how many the child can write and if they are in order.

NATURALISTIC ACTIVITIES

As the child learns that sets and symbols go together, this will be reflected in his daily play activities.

- Mary and Dean have set up a grocery store. Dean has made price tags, and Mary has made play money from construction paper. They have written numerals on each price tag and piece of money. Sam comes up and picks out a box of breakfast cereal and a carton of milk. Dean takes the tags, "That will be four dollars." Sam counts out four play dollar bills. Dean takes a piece of paper from a note pad and writes a "receipt." "Here, Sam."

- Brent has drawn a picture of a birthday cake. There are six candles on the cake and a big numeral "6." This is for my next birthday. I will be six."

- The flannel board and a set of primary cutouts have been left out in a quiet corner. George sits deep in thought as he places the numerals in order and counts out a set of cutouts to go with each numeral.

Each child uses what he has already learned in ways that he has seen adults use these skills and ideas.

INFORMAL ACTIVITIES

The child can work with sets and numerals best through informal experiences. Each child needs a different amount of practice. By making available many materials that the child can work with on his own, the teacher can help each child have the amount of practice he needs. Each child can choose to use the set of materials that he finds the most interesting.

The basic activities for matching symbols to sets and sets to symbols involve the use of the following kinds of materials:[1]

1. Materials where the numerals are 'fixed' and counters are available for making the sets. These are called counting trays and may be made or purchased. They may be set up with the numerals all in one row or in two or more rows.

A.

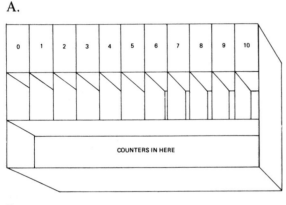

B.

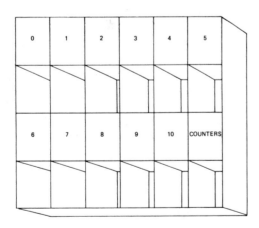

2. Materials where there are movable containers on which the numerals are written and counters of some kind. There might be pennies and banks, cups and buttons, cans and sticks, and similar items.

3. Individual numeral cards with a space for the child to make a set to match.

4. Sets of real things or pictures of things which must be matched to numerals written on cards.

Each child can be shown each new set of materials. He can then have a turn to work with them. If the teacher finds that a child has a hard time, she can give him some help and make sure he takes part in some structured activities.

Informal experiences in which the child writes numerals come up when the child asks how to write his age, phone number, or address. Some children who are interested in writing may copy numerals that they see in the environment — on the clock and calendar or on the numeral cards used in matching and set-making activities. These children should be encouraged and helped if needed. The teacher can make or buy a set of sandpaper numerals. The child can trace these with his finger to get the feel of the shape and the movement needed to make the numeral. Formal writing lessons should not take place until the child's fine muscle coordination is well

Fig. 19-2 "Find this many blocks."

[1] *Workjobs* by Mary Baratta-Lorton (Menlo Park: Addison-Wesley, 1972) is an excellent source for ideas for teacher-made materials in this area.

developed. This means that formal writing should not be taught until a child is five or six years old.

STRUCTURED ACTIVITIES

Structured activities with symbols and sets for the young child are done in the form of games. One type of game has the child match sets and numerals using a theme such as "Letters to the Post Office" or "Fish in The Fish Bowl." A second is the basic "board" game. A third type of game is the Lotto or Bingo type. In each case, the teacher structures the game. However, once the children know the rules, two or more can play on their own. One example of each game is described. With a little imagination, the teacher can think of variations. The last three activities are for the child who can write numerals.

SETS AND SYMBOLS: FISH IN THE FISHBOWL

Objective: To match sets and symbols for the numerals zero through ten

Materials: Sketch eleven fishbowls about 7″ x 10″ on separate pieces of cardboard or posterboard. On each bowl write one of the numerals from zero to ten. Cut out eleven fish, one for each bowl. On each fish, put dots — from zero on the first to ten on the last.

Activity: Play with two or more children. Line up the fishbowls (on a chalk tray is a good place). One at a time, have each child choose a fish, sight unseen. Have him match his fish to its own bowl.

Follow-up:

1. Make fish with other kinds of sets such as stripes or stars.

2. Line up the fish and have the children match the fishbowls to the right fish.

Fig. 19-3 The girls put the fish with the right number of dots in each fishbowl.

SETS AND SYMBOLS:
————BASIC BOARD GAMES————

Objectives: To match sets and symbols

Materials: The basic materials can be purchased or can be made by the teacher. Basic materials would include

- A piece of posterboard (18″ x 36″) for the game board
- Clear Contac® or laminating material
- Marking pens
- Spinner cards, plain 3″ x 5″ file cards, or a di
- Place markers (chips, buttons, or other counters)

Materials for three basic games are shown:

The game boards can be set up with a theme for interest such as the race car game. Themes might be *Going to School, The Road to Happy Land*, or whatever the teacher's imagination can think of.

Activity: The basic activity is the same for each game. Each child picks a marker and puts it on start. Then each in turn spins the spinner (or chooses a card or rolls the di) and moves to the square that matches.

Follow-up:

1. The children can learn to play the games on their own.

2. Make new games with new themes. Make games with more moves and using more numerals and larger sets to match.

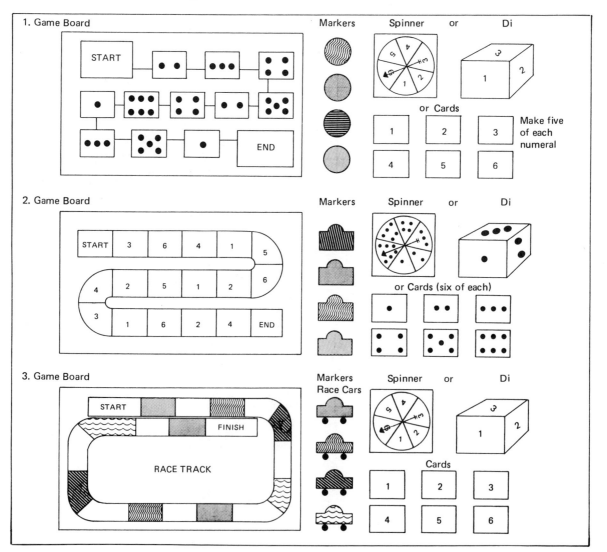

SETS AND SYMBOLS:
——LOTTO AND BINGO GAMES——

Objective: To match sets and symbols

Materials: For both games, there should be six basic game cards each with six or more squares (the more squares, the longer and harder the game). For Lotto, there is one card to match each square. For Bingo, there must be markers to put on the squares, also. For Bingo, squares on the basic game cards are repeated; for Lotto, they are not.

Activities:

1. Lotto Game

   Each child receives a basic game card. The matching cards are shuffled and held up one at a time. The child must call out if the card has his mark on it (dot, circle, triangle) and then match the numeral to the right set. The game can be played until one person fills his card or until everyone does.

2. Bingo Game

   Each child receives a basic game card. He also receives nine chips. The matching set cards are shuffled. They are held up one at a time. The child puts a chip on the numeral that goes with the set on the card. When someone gets a row full in any direction, the game starts again.

Follow-up: More games can be made using different picture sets and adding more squares to the basic game cards. Bingo cards must always have the same odd number of squares up and down and across (3 x 3, 5 x 5, 7 x 7).

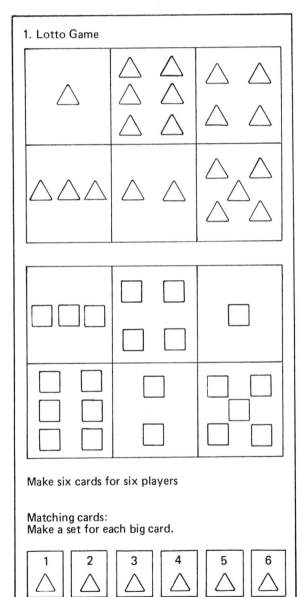

Make six cards for six players

Matching cards:
Make a set for each big card.

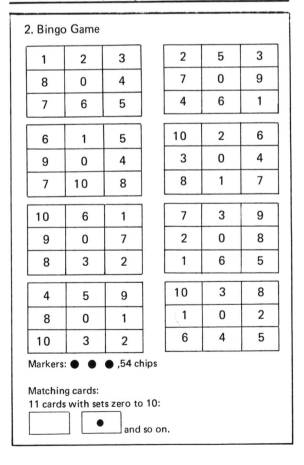

Markers: ● ● ● ,54 chips

Matching cards:
11 cards with sets zero to 10:
and so on.

SETS AND SYMBOLS:
————MY OWN NUMBER BOOK————

Objective: To match sets and symbols

Materials: Booklets made with construction paper covers and several pages made from newsprint or other plain paper, hole puncher and yarn or brads to hold book together, crayons, glue, scissors, and more paper or stickers

Activity: The child writes or asks the teacher to write a numeral on each page of the book. The child then puts a set on each page. Sets can be made using

   a. Any kind of stickers with gummed backs which can be wet and stuck on.

   b. Cutouts made by the child.

   c. Drawings done by the child.

Follow-up: Have the children show their books to each other and then take the books home. Also read to the children some of the number books listed in unit 21.

Fig. 19-4 Dean learns to write numerals.

EVALUATION

With young children most of the evaluation can be done by observing their use of the materials for informal activities. The adult can also notice how the children do when they play the structured games.

For children about to enter first grade, an individual interview should be done using the assessment interviews in the unit and in the Appendix.

SUMMARY

When the child works with sets and symbols, he puts together the skills and ideas learned earlier. He must match, count, classify, compare, order, and associate written numerals with sets.

He learns to match sets to symbols and symbols to sets. He also learns to write each number symbol. The child uses mostly materials which can be used informally on his own. He can also learn from more structured game kinds of activities.

SETS AND SYMBOLS:
—WRITING NUMERALS TO MATCH SETS—

Objective: To write the numeral that goes with a set

Materials: Objects and pictures of objects, chalk and chalkboard, crayons, pencils, and paper

Activity: Show the child some objects or pictures of objects. WRITE THE NUMERAL THAT TELLS HOW MANY _____ THERE ARE. The child then writes the numeral on the chalkboard or on a piece of paper.

Follow-up: Get some clear acetate. Make some set pictures that can be placed in acetate folders for the child to use on his own. Acetate folders are made by taking a piece of cardboard and taping a piece of acetate of the same size on with some plastic tape. The child can write on the acetate with a crayon and then erase his mark with a Kleenex or a soft cloth.

SUGGESTED ACTIVITIES

- Observe children during play. Look for how they match symbols to sets, sets to symbols, or reproduce symbols.

- Add assessment activities in the area of sets and symbols to the Assessment File.

- Make one instructional material which can be used by children during play to help them match symbols to sets or sets to symbols. Share your material with the class. If possible, donate it to a preschool and check to find out if and how they use it.

- Design a structured sets and symbols activity for each of the following:

  a. a game which matches sets and numerals
  b. a basic board game
  c. a Lotto or Bingo game.

  Use one of these with a child or a small group of children. Share your experience with the class.

- Examine catalogs to see what kinds of materials are available that can be used to help children learn sets and symbols.

REVIEW

A. Indicate whether each of the following statements describes a skill the child must have before he goes on to the activities discussed in this unit or a higher level activity.

1. Reproduce numerals
2. Understand that each numeral stands for a certain number of things
3. Match a set to a symbol
4. Match things one-to-one and match sets one-to-one
5. Recognize sets of sizes one to four
6. Count sets of at least ten things without a mistake
7. Recognize each of the numerals from zero to ten
8. Name each of the numerals from zero to ten
9. Understand that each numeral stands for a set of things one more than the numeral before it and one less than the numeral after it
10. Match a symbol to a set
11. Place sets containing different amounts in order from least to most
12. Divide large sets into smaller sets of different amounts

B.   Match each activity in Column II with the correct item in Column I.

I	II
1. To match a symbol to a set	a. writes numerals
2. To reproduce symbols	b. picks three rabbits to go with numeral "3"
3. To match a set to a symbol	c. writes numeral "5" to go with five boats

C.   Indicate **All** the correct answers to each numbered item.

1.   Materials for matching symbols to sets and sets to symbols include

   a. Expensive components
   b. Materials where the numerals are fixed
   c. Materials which can be used to make sets
   d. Materials which include movable containers on which the numerals are written

2.   Young children who are interested in writing numerals

   a. Should not be allowed to do so
   b. Should be advanced to first grade
   c. May copy numerals they see in the environment
   d. Can be given a set of sandpaper numerals

3.   Structured set and symbol activities include

   a. The game Fish in the Fishbowl
   b. Lotto games
   c. Ten pins
   d. Dominoes

4.   Materials to help children reproduce symbols include

   a. Sand
   b. Crayons
   c. Chalk
   d. Paper

5.   Informal sets and symbols activities children do include

   a. Lemonade-selling stands
   b. Bean bag toss
   c. Put the car in the right garage
   d. Place the set card next to the numeral card

unit 20 higher level activities

OBJECTIVES

After studying this unit, the student should be able to

- List the five areas that include higher level math activities
- Describe the three higher levels of classification
- Plan higher level activities for children who are near the stage of concrete operations

The experiences in this unit include

- Further application of skills the child learns through the activities in units 5-19.
- Activities for the child who develops at a fast rate and can do the higher level assessment tasks with ease
- Activities for the older child who needs concrete experiences and variety

The specific areas of experiences are

- Classification
- Shape
- Spatial relations
- Measurement
- Graphs

ASSESSMENT

The teacher looks at the child's level in each area. Then she makes a decision as to when to introduce these activities. When introduced to one child, any one activity could capture the interest of another child who might be at a lower developmental level. Therefore, it is not necessary to wait for all the children to be at the highest level to begin. Children at lower levels can participate in these activities as observers and as contributors. The higher level child can serve as a model for the lower level child. The lower level child might be able to do part of the task following the leadership of the higher level child. For example, if a floor plan of the classroom is being made, the more advanced child might design it while everyone draws a picture of a piece of furniture to put on the floor plan. The more advanced child might get help from the less advanced child when he makes a graph. The lower level child can count and measure while the more advanced child records the results.

CLASSIFICATION

The higher levels of classification are called *multiple classification, inclusion relations,* and *hierarchical classification.* Multiple classification requires the child to classify things in more than one way and to solve matrix problems. Figures 20-1 and 20-2, page 160, illustrate these two types of multiple classification. In figure 20-1, the child is shown three shapes, each in three sizes and in three colors. He is asked to put the ones together that belong together. He is then asked to find another way to divide the shapes.

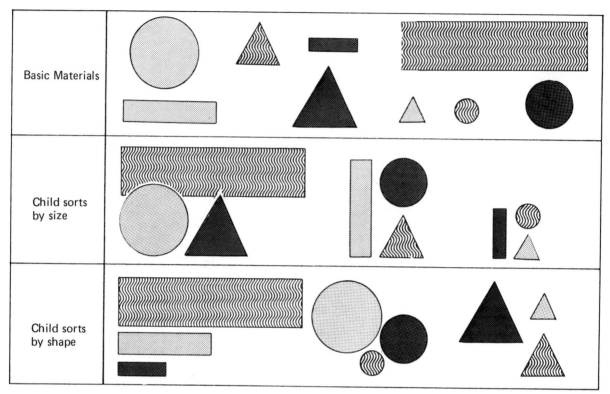

Fig. 20-1 *Multiple classification* **involves sorting one way and then sorting again using different criteria.**

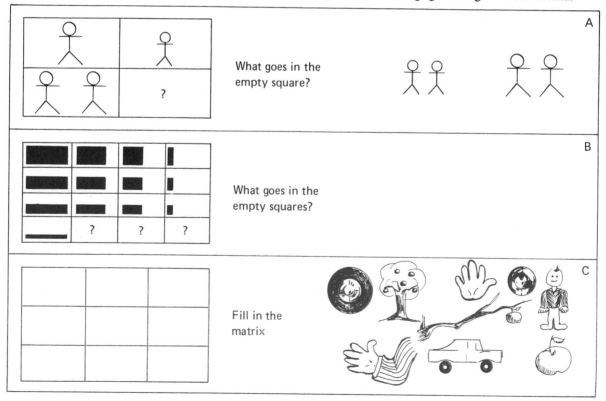

Fig. 20-2 Multiple Classification — Another Type Is the Matrix Problem

The preoperational child will not be able to do this. He centers on his first sort. Some games are suggested which will help the child move to concrete operations.

Matrix problems are illustrated in figure 20-2. A simple two-by-two matrix is shown in the A part. In this case, size and number must both be considered in order to complete the matrix. The problem can be made more difficult by making the matrix larger (there are always the same number of squares in each row, both across and up and down.) The B part shows a four by four matrix. The easiest problem is to fill in part of a matrix. The hardest problem is to fill in a whole blank matrix as is illustrated in the C part.

The preoperational child cannot see that one class may be included within another (class inclusion). For example, the child is shown ten flowers: two roses and eight daisies. The child can divide the flowers into two groups: roses and daisies. He knows that they are all flowers. When asked if there are more flowers or more daisies, he will answer "More daisies." He is fooled by what he sees and centers on the greater number of daisies. He is not able to hold in his mind that daisies

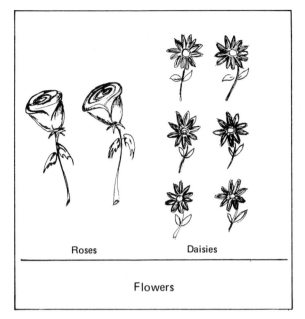

Fig. 20-3 *Class inclusion* **is the idea that one class can be included in another.**

are also flowers. This problem is shown in figure 20-3.

Hierarchical classification has to do with classes being within classes. For example, black kittens → kittens → house cats → cats → mammals. As can be seen in figure 20-4, this forms a *hierarchy,* or a series of ever larger classes.

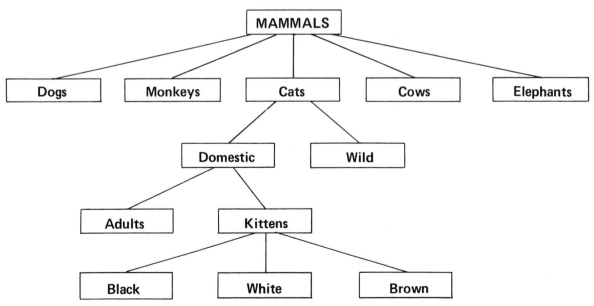

Fig. 20-4 In a *hierarchical classification***, all things in each lower class are included in the next higher class.**

The following activities will help the transitional child (usually age five to seven) to enter concrete operations.

HIGHER LEVEL CLASSIFICATION: MULTIPLE CLASSIFICATION, ──────RECLASSIFY──────

Objectives: To help the transitional child learn that groups of objects or pictures can sometimes be sorted in more than one way

Materials: Any group of objects or pictures of objects which can be classified in more than one way: for example, pictures or cardboard cutouts of dogs of different colors (brown and white), sizes (large and small), and hair lengths (long and short)

Activity: Place the dogs in front of the child. WHICH DOGS BELONG TOGETHER or ARE THEY THE SAME? Note whether he groups by size, color, or hair length. NOW, WHAT IS ANOTHER WAY TO PUT THEM IN SETS?. . . .CAN THEY BE PUT LIKE (name another way)? Put them in one pile again if the child is puzzled. OKAY, NOW TRY TO SORT THE _____ FROM THE _____ . Repeat using different criteria each time.

Follow-up: Make other sets of materials. Set them up in boxes where the child can get them out and use them during free play time. Make some felt pieces to use on the flannel board.

HIGHER LEVEL CLASSIFICATION: MULTIPLE CLASSIFICATION, ──────MATRICES──────

Objectives: To help the transitional child see that things may be related on more than one criteria

Materials: Purchase or make a matrix game. Start with a two-by-two matrix and gradually increase the size (three-by-three, four-by-four, and so on). Use any of the criteria from unit 7 such as color, size, shape, material, pattern, texture, function, association, class name, common feature, or number. Make a game board from posterboard or wood. Draw or paint permanent lines. Use a flannel board, make the lines for the matrix with lengths of yarn. An example of a three-by-three board is shown.

Activities: Start with the matrix filled except for one space and ask the child to choose from two items the one that goes in the empty space. WHICH ONE OF THESE GOES HERE? After the item is placed: WHY DOES IT BELONG THERE? Once the child understands the task, more spaces can be left empty until it is left for the child to fill in the whole matrix.

Follow-up: Add more games which use different categories and larger matrices.

Items to be placed:

1. Color and size: red, green and yellow apples:

2. Position and class: dog, cat, bird; rightside-up, upside-down and sideways.

3. Number and shape: triangles, squares, circles; rows of one, two, and three.

HIGHER LEVEL CLASSIFICATION:
————CLASS INCLUSION————

Objectives: To help the transitional child see that a smaller set may be included within a larger set

Materials: Seven animals. Two kinds should be included (such as horses and cows, pigs and chickens, dogs and cats). There should be four of one animal and three of the other. These can be cutouts or toy animals.

Activity: Place the animals within an enclosure (a yarn circle or a fence made of blocks). WHO IS INSIDE THE FENCE? Children will answer "horses," "cows," "animals." SHOW ME WHICH ONES ARE HORSES (COWS, ANIMALS). ARE THERE MORE HORSES OR MORE ANIMALS? . . . .HOW DO YOU KNOW? . . .LET'S CHECK (use one-to-one correspondence).

Follow-up: Play the same game. Use other categories such as plants, types of material, size, and so on. Increase the size of the sets.

HIGHER LEVEL CLASSIFICATION:
————HIERARCHICAL————

Objective: To help the transitional child see that each thing may be part of a larger category (or set of things)

Materials: Make some sets of sorting cards. Glue pictures from catalogs and/or workbooks onto file cards or posterboard. The following are some that can be used:

1. one black cat, several house cats, cats of other colors, one tiger, one lion, one panther, one bobcat, one dog, one horse, one cow, one squirrel, one bear

2. one duck, three swans, five other birds, five other animals

3. one teaspoon, two soup spoons, a serving spoon, two baby spoons, three forks, two knives

Activities: Place the cards where they can all be seen. Give the following instructions:

1. FIND ALL THE ANIMALS. . . .FIND ALL THE CATS. . . .FIND ALL THE HOUSE CATS . . . .FIND ALL THE BLACK CATS. Mix up

Fig. 20-5 Other categories, such as size, may be used in follow-up activities.

the cards and lay them out again. PUT THEM IN SETS THE WAY YOU THINK THEY SHOULD BE. When the child is done, WHY DID YOU PUT THEM THAT WAY? Mix them up and lay them out. IF ALL THE ANIMALS WERE HUNGRY, WOULD THE BLACK CAT BE HUNGRY? IF THE BLACK CAT IS HUNGRY ARE ALL THE ANIMALS HUNGRY?

2. FIND ALL THE ANIMALS. . . .FIND ALL THE BIRDS. . . . . FIND THE WATER BIRDS . . . . FIND THE DUCK. Mix up the cards and lay them out again. PUT THEM IN SETS THE WAY YOU THINK THEY SHOULD BE. When the child is done, WHY DO THEY BELONG THAT WAY? Mix them up and lay them out again. IF ALL THE BIRDS WERE COLD WOULD THE DUCK BE COLD? IF THE DUCK WAS COLD WOULD ALL THE WATER BIRDS BE COLD? IF ALL THE ANIMALS WERE COLD WOULD THE WATER BIRDS BE COLD?

3. FIND ALL THE THINGS THAT WE EAT WITH. . .FIND ALL THE KNIVES. . . .FIND ALL THE FORKS. . . .FIND ALL THE SPOONS. Mix them up and lay them out again. PUT THEM IN SETS THE WAY YOU THINK THEY BELONG. When the child is done, WHY DO THEY BELONG THAT WAY? Mix them up and lay them out again. IF ALL THE SPOONS ARE DIRTY WOULD THE TEASPOON BE DIRTY?. . .IF ALL THE THINGS WE EAT WITH WERE DIRTY WOULD THE BIG SPOON BE DIRTY?. . .IF THE TEASPOON IS DIRTY, ARE ALL THE OTHER THINGS WE EAT WITH DIRTY TOO?

Follow-up: Make up other hierarchies. Leave the card sets out for the children to sort during free play. Ask them some of the same kinds of questions informally.

SHAPE

Once the child can match, sort, and name shapes he can also reproduce shapes. This can be done informally. The following are some materials that can be used.

Geoboards can be purchased or made. A *geoboard* is a square board with nails (heads up) or pegs sticking up at equal intervals. The child is given a supply of rubber bands and can experiment in making shapes by stretching the rubber bands between the nails.

A container of pipe cleaners or straws can be put out. The children can be asked to make as many different shapes as they can. These can be glued onto construction paper. Strips of paper, toothpicks, string, and yarn can also be used to make shapes.

SPATIAL RELATIONS

The child can learn more about space by reproducing the space around him as a floor plan or a map. Start with the classroom the child is in for the first map. Then move to the whole building and finally to the neighborhood and the town or city.

HIGHER LEVEL ACTIVITIES: —SPATIAL RELATIONS, FLOOR PLANS—

Objective: To relate position in space to symbols of position in space

Materials: Large piece of posterboard or heavy paper; markers, pens, construction paper, glue, crayons, scissors; some simple sample floor plans

Activity:

1. Show the children some floor plans. WHAT ARE THESE? WHAT ARE THEY FOR?. . .IF WE MAKE A FLOOR PLAN OF OUR ROOM WHAT WOULD WE PUT ON IT? Make a list.

2. Show the children a large piece of posterboard or heavy paper. WE CAN MAKE A PLAN OF OUR ROOM ON HERE. EACH OF YOU CAN MAKE SOMETHING THAT IS IN THE ROOM. JUST LIKE ON OUR LIST. THEN YOU CAN GLUE IT IN THE RIGHT PLACE. I'VE MARKED IN THE DOORS AND WINDOWS FOR YOU. As each child draws and cuts out an item (a table, shelf, sink, chair) have him show you where it belongs on the plan and glue it on.

Follow-up: After the plan is done, it should be left up on the wall so the children can look at it and talk about it. They can also add more things to the plan. The same procedure can later be used to make a plan of the building. Teacher and children should walk around the whole place. They should talk about which rooms are next to each other, which rooms are across from each other. Sticks or straws can be used to lay out the plan.

HIGHER LEVEL ACTIVITIES: ——SPATIAL RELATIONS, MAPS ——

Objective: To relate position in space to symbols of position in space

Materials: Map of the city, large piece of posterboard or heavy paper, marking pens, construction paper, glue, crayons, scissors

Activity: Show the children the map of the city (or county in a rural area). Explain that this is a picture of where the streets would be if the children were looking down from a plane or a helicopter. Mark each child's home on the map with a small label with his name. Mark where the school is. Talk about who lives closest and who lives farthest away. Each child's address can be printed on a card and reviewed with him each day. The teacher can help mark out the streets and roads. The children can then cut out and glue down strips of black paper for the streets (and/or roads). Each child can draw a picture of his home and glue it on the map. The map can be kept up on the wall for children to look at and talk about. As field trips are taken during the year, each place can be added to the map.

Follow-up: Encourage the children to look at and talk about the map. Help them add new points of interest. Help children who would like to make their own maps. Bring in maps of the state, the country, and the world. Try to purchase United States and world map puzzles.

MEASUREMENT

Most young children will not get into formal measurement. That is, they will not measure in feet, yards, meters, kilometers, pounds, liters, and other standard units. However, some may become interested. The teacher should encourage this interest. Therefore, even teachers of young children need to know about the process of changing to the metric system of measure that is taking place. During this transitional stage, the teacher will have to decide whether to use the present English units first and then introduce the metric system or to start the child right off with metrics.

Each class should be equipped with a set of metric measuring devices: a celsius thermometer, a set of metric graduated beakers, a meter stick, a set of metric weights for the pan balance, and a set of metric interlocking cubes. The teacher should become familiar with the metric units: centimeters, decimeters, and meters for shorter lengths; decameter, hectometer and kilometer for longer lengths; liters, deciliters, and centiliters for volume; kilogram, gram, and milligram for weight; degrees celsius for temperature. Figure 23-5 in unit 23 illustrates metric units.)

GRAPHS

The fourth level of graphs introduces the use of squared paper. The child may graph the same kind of things as discussed in unit 16. He will now use squared paper with squares that can be colored in. These should be introduced only after the child has had many experiences of the kinds described in unit 16. The squares should be large. A completed graph might look like the one shown in figure 20-6, page 166.

SUMMARY

Experiences with some higher level activities in the areas of classification, shape, spatial relations, measurement, and graphs are included. These activities are for children who are in the transitional stage of development which leads to concrete operations.

SUGGESTED ACTIVITIES

- Review children's magazines. Make a list of the activities suggested that relate to helping children learn math symbols. Share these with your class.

- Observe kindergarten age children. What evidence was noted that some children were ready for and some were using higher level math activities? Share what you observed with your class.

- Interview a kindergarten teacher. Ask how she plans math activities for children who are progressing towards the concrete operational stage.

- Select, prepare, and present two higher level math activities for transitional children. Share these with your class.

- Add higher level math activities to your Activities File.

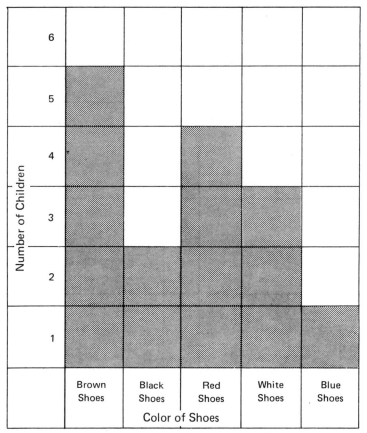

Fig. 20-6 Squared paper graphs can be made by the child who has had many experiences with simpler graphs.

REVIEW

A.   Name the five areas that include higher level math activities.

B.   Describe the three higher levels of classification.

C.   Match the terms in Column II with the correct item in Column I.

I	II
1. Apple, fruit, food	a. Multiple classification
2. Matrices	b. Class inclusion
3. Two balls, five toy cars — are there more cars or more toys?	c. Hierarchical classification
4. Shape, size, and color	
5. Pencils, crayons, chalk, writing implements	
6. Three carrots and two stalks of celery — are there more carrots or more vegetables?	

D. Given the following situations, tell what higher level activity the child seems interested in and how the teacher should respond.

1. Mary says, "Susie lives across the street from me."

2. Jim takes the box of shape attribute blocks and begins to stack them into different piles.

3. Karen takes some plastic fruit and wooden people and places them on the balance scale.

4. Alissa points to the picture in the story book and says, "See the round apple."

5. Alissa and Jim are playing. Alissa says, "Blue is my favorite color." Jim says he likes yellow best. He says, "More kids like yellow than blue." Alissa says, "No, they don't."

Section 5 The Math Environment

unit 21 materials and resources

OBJECTIVES

After studying this unit, the student should be able to

- Categorize math materials as manipulative, pictorial, or abstract
- Set up a math learning center
- Select appropriate math materials applying educational criteria
- List examples of basic math materials for specific math skill development

A variety of materials is important to help children learn math. The teacher can select materials that are easy to find, are safe, and can be used in many ways. Commercial materials and math kits can be purchased. However, math materials do not have to be expensive to be useful. Many math materials are items found in the home or available free from a local resource.

Math materials can be categorized as manipulative (concrete), pictorial (semiconcrete), and abstract (symbolic). Inexpensive manipulative math materials include buttons, egg cartons, blocks, and balls of different sizes. Most of the math materials described in this unit are manipulative.

Pictorial materials include a collection of children's books, pictures, and poems. These

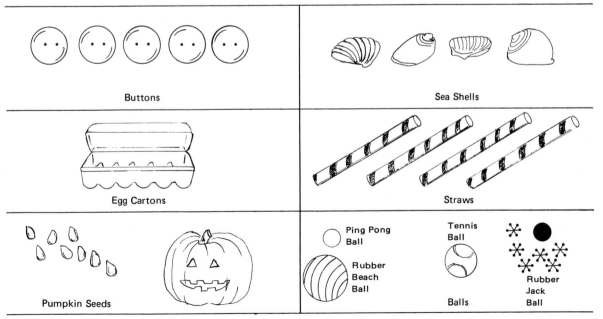

Fig. 21-1 Inexpensive Math Materials

help teach math vocabulary. They also illustrate the use of math in a variety of settings. They expand children's ideas of how math can be used. Since there are so many pictorial math resources, each should be selected carefully. The teacher should ask, "What is the main math idea in this story?" "How will it help Jim, Betty, or Dan understand the math idea better?" The books should be artistic, colorful, and well written. For example, an excellent children's book that includes math concepts is *Over in The Meadow* by Ezra Jack Keats. The rhythm of

> "Over in the meadow, in the sand, in the sun,
> Lived an old mother turtle and her little turtle
>   one.
> "Dig!" said the mother.
> "I dig," said the one.
> So he dug all day,
> In the sand, in the sun"

is appreciated by the teachers as well as the children. Appendix B contains a list of math books suitable for young children.

Abstract materials are symbolic figures such as dots and numerals. Flannel boards and flannel board figures can be used for abstract math experiences. For example, Miss Thom is not sure if Tammy understands sets matched to numerals. She cuts from flannel the numerals 0-9. She cuts out forty-five pumpkins. On the flannel board Miss Thom puts the numerals 1, 2, 3, and 4. She gives Tammy the forty-five pumpkins. She tells Tammy to put the proper number of pumpkins by the right symbol. If Tammy can do this, Miss Thom can further check to see if she knows all the nine numerals. Abstract materials are used only when a child has had many experiences with concrete materials. The most abstract activity would involve some reading and writing skills.

MATH LEARNING CENTER

Math materials should be displayed so that children can easily see them, use them,

Fig. 21-2 Open containers and low shelves invite children to use the math materials.

and put them away. Too many math materials stacked in an unorganized pile limits their use. It is better to place the materials side by side in baskets or other open containers on low shelves in the math learning center. This tells the child that these materials are available and can be used. This helps children learn proper replacement locations.

Teachers should change some of the math materials according to the needs of individual children. This may be done on a daily, weekly, or monthly basis. Rotating special materials is called creating a responsive environment. A responsive environment is one that continually challenges and interests children. A teacher should be assigned to the math learning center. She can observe which math materials are being used. She can observe which children use each kind of material. The teacher can encourage children to use new math materials. She does this by sitting down and playing a new math game with a child. This adult-child interaction is often the stimulus for child-child interaction. After the adult shows the child how to play a game, the child then teaches a friend to play it.

SELECTING MATH MATERIALS

Each child has a variety of needs and potential within his math development. Each school usually includes children who are at various levels of math development. It is important to provide a wide array of math materials. Children need to play freely with these materials.

Early math materials are those that provide firsthand sensory experiences for touching, tasting, hearing, and seeing. The importance of allowing young children to explore materials in a richly provided environment cannot be overemphasized. Exercise of small muscles and fine motor control occurs with peg boards, bead stringing, tinker toys, sequence cards, and stack clowns.

In choosing any math material, an adult should ask the following questions:

* Is it designed for young children?
* Is it economical?
* Is it safe?
* Is it durable?
* Is it versatile?
* Is it easily supervised?

Materials for young children should be stimulating. This is facilitated through appropriate size, height, and weight of the equipment. They should be "child-size." The cost of a material is related to its use. Is the material a special piece of equipment used only once a year? Or is the material one that is used daily? Blocks, for example, are used daily by children. All equipment must be safe for children to use. Hazardous features, such as sharp or pointed edges that could injure a child, should be avoided. Painted surfaces should have a nontoxic and nonflammable paint. Paper materials should be laminated to strengthen them. Wooden items are desirable as long as they have smooth edges. Metal materials can rust. Plastic items can crack. Quality math materials should serve more than one purpose. For example, blocks can be used in many ways. Placement of materials is an important

Fig. 21-3 Children help each other learn.

consideration for supervision. Climbing bars in the outdoor play space should be located where there is room for the children to explore.

Before deciding what new math materials to obtain, the teacher lists what is already available. She should make notes as to the condition of current math materials. Are Lego pieces missing? Do games need mending? After this inventory, a list is made of the most needed materials to collect or purchase first. The teacher now considers the children's needs, both individually and as a group. She also considers the indoor and outdoor learning environment and the amount of room available. She considers the math objectives of the program. She considers how the materials will be used. Are they for naturalistic, informal, or structured activities?

After these decisions are made, the teacher is ready to purchase the math materials or explore an alternate source for obtaining them. One alternate source is to make the math material. For example, beanbags are easy to make. Scrap materials may be used as shown in figure 21-4, page 172. Another source is to find people who are willing to donate some materials. Parents usually are willing to save egg cartons, buttons, or seeds for the math program. Lumber companies often donate scrap materials.

1. Use scrap pieces of felt fabric or other sturdy material in bright primary colors of red, yellow, and blue.

2. Cut 2 each of the 4 basic shapes. Make the square 4″ x 4″. The circle, rectangle, and triangle should be similar in size.

3. Cut small black felt numerals 1, 2, 3, and 4. Cut small black felt shapes in numbers to match each of the number symbols (1 circle, 2 rectangles, 3 triangles, and 4 squares, for example).

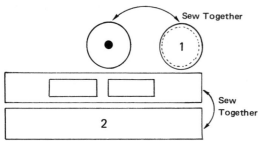

4. Machine sew one number symbol on one of each of the 4 basic shapes that were cut from felt earlier. Sew all the similar small felt shapes to each of the remaining 4 basic shapes, as shown.

5. Machine sew the matching shape pieces together. Stitch 1/4 inch from edge. Leave 1-inch opening. Stuff the bags with sawdust, cutup nylon hosiery pieces, or commercial pillow stuffings. Machine or hand sew the opening.

Fig. 21-4 Number and Shape Beanbags

When the needs are clear and decisions have been made as to which math items can only be obtained by buying them new, the teacher looks through catalogs or visits local school supply stores. There are many quality resources. It is important to look at several catalogs. Although supply companies offer similar items, each company also has special math items. Some well known catalogs are listed in figure 21-5.

BASIC MATH MATERIALS

Basic math materials can be grouped according to their major use. However, many are used for multiple purposes. Math learnings can be acquired by children as they play. Children will develop math skills as they play freely.

Matching (unit 5) is the understanding that one group has the same number of things as another. Matching materials include wood inlay puzzles, play tiles, pegboards and pegs, picture lotto games, dominoes, fitting sets, snap beads, snap brads, wooden beads, and strings. Also, very useful are colored inch cube blocks. A self-learning aid called the Tutogram® can be used for matching. A card

is placed in the machine. When the child inserts the rod in the proper match, the top knob lights and the machine buzzes. The child instantly knows when he has made the correct match.

Counting (unit 6) includes rote and rational counting. Rote counting involves reciting the names of the numerals in order from memory. Rational counting involves attaching the name of each numeral in order to a series of objects in a group. Almost anything can be counted. Crayons, marbles, poker chips, sticks of all kinds, buttons, pegs, unifix cubes, shells, money, toothpicks, beads, dried beans or peas, popcorn kernels, small toys, counting frames, and inch cube blocks are examples. Flannel boards and flannel shapes or objects may be used for rational counting activities. Most of the above counting objects can be counted out in egg cartons or other containers that are marked with symbols.

Classifying (unit 7) consists of sorting and grouping objects. Accessories that children like to classify are small farm animals, zoo animals, people, fruit, flowers, and play dishes. These can be made of wood, metal,

- BBT Learning Materials; 1515 Broadway, New York, New York 10036
- Bowmar; 622 Rodier Drive, Glendale, California 91201
- Cabdev, Incorporated; 31 Progress Ave. — Unit 9 — Scarborough, Ontario, MIP 456 Canada
- Childcraft; 20 Kilmer Road, Edison, New Jersey 08817
- Community Playthings; Rifton, New York 12471
- Creative Publications; P.O. Box 10328, 3977 East Bayshore Road, Palo Alto, California 94303
- Developmental Learning Materials; 7440 Natchez Ave., Niles, Illinois 60648
- EDcom Systems, Inc.; Princeton, New Jersey 08540
- Educational Teaching Aids (ETA); 159 West Kinzie Street, Chicago, Illinois 60648
- Ideal School Supply Company; 11000 South Lavergne Avenue, Oak Lawn, Illinois 60453 (Early Learning Special Education)
- Judy Co.; 310 N. Second St., Minneapolis, Minnesota 55402
- Learning Games, Inc.; 34 South Broadway, White Plains, New York 10601
- Novo Educational Toys and Equipment Corp.; 124 West 24th Street, New York, New York 10011
- Teaching Resources, 100 Boylston Street, Boston, Massachusetts 02116

Fig. 21-5 Math Equipment Catalogs

Fig. 21-6 "Oh dear, will I get the right match?"

Fig. 21-7 "Yeah, I did it."

Fig. 21-8 "Teacher, these are wooden people."

rubber, or plastic. Buttons, hardboard geometric shapes, felt shapes, wooden beads, unit blocks, and color cubes also are good for classifying.

Comparing (unit 8) is finding a relationship between two things or sets of things. All of the objects used for classifying can be used to make comparisons. Hoops or loops made from shoelaces, ribbon, rope, or string can be used to sort sets. Paper plates, pieces of construction paper, muffin tins, or boxes are also used to separate items.

Each thing the child meets in the environment has *shape* (unit 9). Many commercial shape materials can be purchased. Shape sorting boxes and shape coordination boards are made by several companies. Clear shape stencils and size and shape puzzles are also available. Montessori materials include well made shape cylinders and insets.

Space as a part of math and geometry was discussed in unit 10. Construction

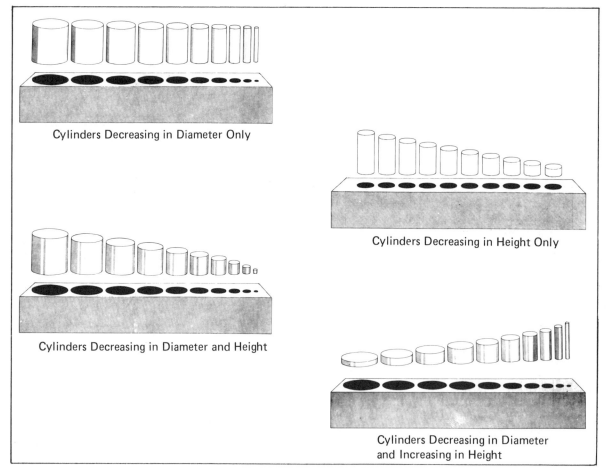

Cylinders Decreasing in Diameter Only

Cylinders Decreasing in Height Only

Cylinders Decreasing in Diameter and Height

Cylinders Decreasing in Diameter and Increasing in Height

Fig. 21-9 Montessori Shape Cylinders and Insets

materials help a child learn about space. Position, direction, and distance are manipulated by children when they build. Besides the basic unit blocks, which are discussed in unit 22, there are many commercial construction sets for children. Space is also explored through organization and pattern materials. Some of these include geoboards, parquetry blocks, hexangle pattern set Peg-A-Pattern®, grid mosaic, play tiles, sewing boards, stringing rings, wooden beads and strings, or see-thru threading shapes.

The concept of *parts and wholes* is basic for learning about fractions (unit 11). Many commercial companies produce lotto cards and puzzles that children can manipulate to learn about parts and wholes. Naturalistic experiences such as food preparation also help children understand this concept.

Ordering involves comparing more than two things or more than two sets. Ordering was discussed in unit 13. There are many commercial ordering items. Nesting boxes, dolls, and boats are available. Color stacking discs, ring-a-rounds, a learning tower, pan pile-cups, handy boxes, and pagoda tower builder can be purchased. Home items are measuring cups and spoons as well as different sizes of containers.

Fig. 21-10 **Building with plastic bricks is challenging.**

Measurement was discussed in units 14-15. Scales, tape measures, yardsticks, and rulers should be available at the math center. Liter measuring pitchers, nesting graduated beakers, and thermometers are also appropriate math tools. Many of these items along with funnels and sieves should be available at the water and sand table areas. Sifters, wooden

Blockbusters	Lincoln Logs	Play Squares	Beam and Boards
Lego	Baufix	Octons	Habitat
Free Form Posts	Crystal Climbers	Girders	Busy Blocks
Sprocketeers	Rig-A-Jig	Play-Panels	Space Wheels
Tinkertoys	Color Cone	Wonderforms	Magnastiks
Toy Makers	Tectonic	Geo-D-Stix	Keeptacks
Cloth Cubes	Structo-Brics	Connector	Balancing H Blocks
Snap Wall	Giant Structo-Cubes	Wood'n Molds	Block Head
Lock & Stack Blocks	Floresco	Poki Blocks	Snap-N-Play Blocks
Giant Interlockers	Ring-A-Majigs	Multi-fit	Channel Blocks

Fig. 21-11 **Construction Materials for Math (Blocks unit excluded.)**

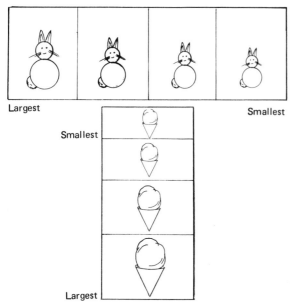

Largest   Smallest

Smallest

Largest

Fig. 21-12 Which is largest? Which is smallest?

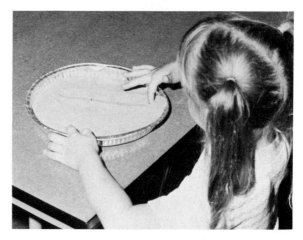

Fig. 21-13 Writing the numeral "one" in sand is a sensory experience.

spoons, plastic pails, strainers, and assorted pans should be at the sand table. Water toys that enhance math development include plastic funnels, pitchers, cups, and bottles. Time and sequence materials include a day by day calendar, clocks, timers, hourglass, and sundial. Sequence cards and seasons charts may be made or purchased.

Additional materials include play money, a cash register, spinners, and dice. To help. young children learn numerals, there are many games and puzzles. Tactile numeral teaching aids are also available. Precut numbers and symbol cards may be used for many different activities. Colored sand can be used for finger number printing. Posterboard, marking pens, glue, construction paper, scissors, and crayons are needed to make graphs, maps, and charts.

SUMMARY

Math materials may be classified as manipulative, pictorial, and abstract. The math learning center is responsive to the children's needs. Many decisions must be made before math materials are obtained. Math materials may be purchased, may be homemade, and may include a collection of free objects. Basic materials for each unit are listed.

SUGGESTED ACTIVITIES

- Visit a local toy supply company. Make a list of available math materials and their prices. Bring the list to class.

- Participate in a small group in class and compare math materials lists. Make cooperative decisions as to what math materials should be purchased if a new center is to be opened. Each group should consider cost, as well as use.

- Go to the library. Find and read at least ten children's picture books that contain math ideas. Write a description of each one. Tell how each book could be used with children.

- Make two different "homemade" math resources that could be used with young children. Share the materials with the class. Be prepared to show the class how the resources can be made.

- Visit two preschools or kindergarten classes. Ask to look at the math materials used. Diagram the math learning center or math shelves. Share the information with your class. Tell the strong and weak points of the math setup. List any changes which would make it better.

- Add one file card that lists math materials to each of the activity units in your file.

- Send for free commercial catalogs. Make a list of math materials that are new in each. Write down the descriptions presented in the catalogs. Share this information with your class.

REVIEW

A. Describe each of the following.

1. Manipulative math materials

2. Pictorial math materials

3. Abstract math materials

B. Respond to each of the following situations.

1. Teacher A has 6 lotto games, a geoboard with no rubber bands, and some odds and ends of blocks on a table in the corner of the nursery school. Teacher B has 2 lotto games, some buttons in a container, a stack clown, a wooden shape puzzle, and blocks. Which teacher has a better selection of math materials? Why?

2. Teacher C says she does not believe in teaching math to young children. Her preschool has a sand table, water table, and a woodworking set. Children are observed playing in the sand with measuring cups and bowls. Some children are also weighing plastic fruit on a balance scale in the housekeeping center. Is Teacher C teaching math? If you think so, explain how she really is teaching math to children.

3. Teacher D has been given fifty dollars to update her math equipment. She asks for help. What advice could be given?

4. Teacher E has just read this book. She is trying very hard to provide a responsive math environment. However, she does not know how to begin. What suggestions can you give her?

C. Select appropriate math materials for the following.

1. A preschool center that already has picture books, pegboard and pegs, wooden beads and strings, unifix cubes, assorted blocks, a learning tower, and a woodworking area.

2. A kindergarten that already has picture books and ditto materials.

D. Match the correct items from Column II with each item in Column I.

I	II
1. Thermometer	a. Matching
2. Color stacking discs	b. Number and counting
3. Connector	c. Sets and classifying
4. Dominoes	d. Comparing
5. Marbles	e. Shape
6. Plastic flowers	f. Space
7. Parquetry blocks	g. Parts and whole
8. Scale	h. Ordering
9. Calendar	i. Measurement
10. Attribute shapes	
11. Boxes of different sizes	
12. Pictures of people in various jobs	
13. Big/small blocks	
14. Half a sandwich	
15. Lego	

unit 22 math in action

OBJECTIVES

After studying this unit, the student should be able to

- Plan and use woodworking for math experiences
- Plan and use blocks for math experiences
- Plan and use math games for math experiences
- Share math fingerplays and songs with children

In nursery school, day care, and kindergarten, math goes on all the time. The young carpenter measures wood. The block builder engineers a building. This is math in action. Throughout the day the teacher of young children may use fingerplays and songs that contain math ideas. Many games, such as musical chairs, that young children like to play help them understand the use and the meaning of math.

WOODWORKING

Both boys and girls like to work with wood. It provides concrete experiences with measurement, balance, and spatial relationships. As children work with wood, they compare the size of one piece of wood to another. They learn to judge length, width, and thickness. Their conversations contain phrases such as "long enough," "about the same," and "I need more."

Effective woodworking requires a sturdy workbench, real tools of good quality, and an assortment of soft wood. The workbench should be large enough for two or three children to build at one time. It must be put in a place where it can be well supervised. Work surfaces can be a purchased workbench designed for young children, an old sturdy table at children's height, or a large tree stump.

High quality tools should be purchased for woodworking. Figure 22-1 illustrates the basic carpentry tools for children ages four and five. The tools should be located where they are accessible and easy to keep in order. Tools should not be allowed to get rusty. Other accessories include short nails with large heads, coarse sandpaper, white glue, and wire.

A soft wood such as pine makes a good surface to pound. For the child who is learning to use the saw, lengths of two inches by one-half inch should be provided. The pine

Two 7 oz.
Claw Hammers

One 12"
Back-type Saw

One 3"
Screwdriver
One 4"
Screwdriver

One Combination
Pliers

Fig. 22-1 Basic Woodworking Tools for Four/Five-Year-Olds

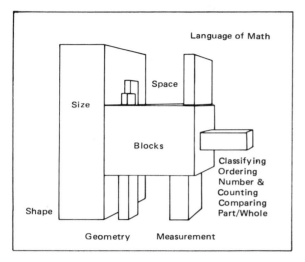

Fig. 22-2 Children learn from blocks.

wood should be placed in a vise, so that it will be held steady while the child saws it.

Lumberyards often give away scraps of wood. Styrofoam, chunks of layered corrugated cardboard, or acoustical ceiling tile also can be used for carpentry experiences.

To help children explore space creatively, the adult should also provide containers of odds and ends. Bottle caps can be wheels. Straws can be cut and used for oars. Plastic lids can be windows. Sponges, spools, cloth scraps, ribbons, and bows help the young woodworkers create all sorts of objects. Even feathers, flowers, and large weeds can be nailed or glued onto a piece of wood. Children get much satisfaction from their woodworking projects. In creating them, the children make plans and sequence their activity. When the project is finished, the children want to talk about what they have done. This is the time the alert adult can help a child use his math vocabulary.

BLOCKS

Blocks are probably the most used play material for preschool programs and kindergarten. Basic math concepts are developed as children explore the relationship of unit block sizes and shapes. Math terms such as "one more" or "two wide" are used by children as they experiment with blocks. Likenesses and differences in form can be seen as well as how forms fit together. When children use blocks, they are working in space. They learn to manipulate length, width, and height. They measure. They discover that a certain number of blocks are required to equal the length or width of another block. As children take two half-circle blocks and put them together, they make a circle. They learn about parts and wholes. They learn about fractions.

The block area should be located where there is adequate floor space for building and away from other activities. Low, open shelved block cabinets should be provided so that children can easily reach the blocks and put them away. Shelves should be marked to help children store blocks. The teacher's role is to gradually introduce more blocks or a variety of shapes as children's needs grow. The teacher can ask questions to help children discover how things are different or how they are alike. For example, Tommy and Sally are each building structures. Miss Raha says, "Sally, why is your house two blocks wider than Tommy's?"

Blocks are available by sets. Each set is an assortment of shapes and sizes. Unit blocks are the basic forms. The basic unit is a brick-shaped rectangle (1 3/8″ x 2 3/4″ x 5 1/2″).

Fig. 22-3 Marked shape pieces help children store blocks properly.

Figure 22-4 lists the great variety of shapes combining straight and round surfaces.

Unit blocks should be of durable hardwood with all edges beveled to prevent wear and splintering. Smoothly sanded surfaces are important for safe handling. Precise dimensions are important for effective building.

	Name	Nursery	Kgn. & Primary
	Square	40	80
	Unit	96	192
	Double Unit	48	96
	Quadruple Unit	16	32
	Pillar	24	48
	Half Pillar	24	48
	Small Triangle	24	48
	Large Triangle	24	48
	Small Column	16	32
	Large Column	8	16
	Ramp	16	32
	Ellipse		8
	Curve	8	16
	¼ Circle		8
	Large Switch & Gothic Door		4
	Small Switch		4
	Large Buttress		4
	½ Arch & Small Buttress		4
	Arch & ½ Circle		4
	Roofboard		24
	Number of Shapes	12	23
	Number of Pieces	344	760

Fig. 22-4 Child Craft Block Sets

Blocks are expensive. However, with proper care, blocks should last at least ten years. They should be kept dry and free of dust. If oil or wax is put on occasionally, the blocks will last longer. There are toys which may be used with blocks. Cars, trucks, farm and zoo animals, block people, and colored inch cubes add stimulation and variety to block play.

Teachers of young children have observed that all children pass through stages in block construction, figure 22-5, page 182. When first introduced to blocks, a child carries the blocks from place to place. He handles them, but does not build with them. Second, the child makes rows. Third, he makes bridges. He connects two blocks with a third. In the fourth stage, the child places blocks in such a way that they enclose a space. This is called making enclosures. Fifth, the child uses the blocks to make decorative patterns. Much symmetry (balance) is observed. In the sixth stage, the child names his structures and dramatic play begins. Finally, the child makes buildings that represent actual structures he knows. This may be his school or his house or apartment building. After they reach this stage, children use their structures for dramatic play.

There are other kinds of building materials. Large hollow wood blocks have been used for a long time in preschools. There are also colorful plastic substitutes and corrugated board boxes available at a lower cost. Cardboard boxes are also used imaginatively by children. As they walk around, climb over, crawl into, and peek in and out of boxes, children learn math vocabulary. All of these building materials show differences in weight, size, shape, and texture. As children pick them up, move them, and build with them, they develop an awareness of comparisons between size, proportion, and volume. They have the opportunity to manipulate their environment. Math is in action.

A. Child handles blocks.

B. Child builds rows.

C. Child connects blocks.

D. Child builds enclosures.

E. Child makes decorative patterns.

F. Child names his structure. "Look, I made a fire station."

Fig. 22-5 Stages in Block Construction

Fig. 22-6 In which stage of block construction are these children?

MATH GAMES

Games become a part of a child's experience at an early age. Mother plays "peek-a-boo" with the infant. As he gets older, his brothers and sisters play "going-in-the-car" and "let's find" games. As his father tosses him and catches him, the child experiences free space. Grandmother plays "how big" is Tommie today. By the time the child goes to school, he has played many informal games and has used his senses in them.

Games in the preschool should not be competitive. Young children do not understand competition. Games should be simple and contain only two or three directions. The teacher presents games with certain aims in mind to aid in understanding math ideas.

Board games provide an excellent way to teach math. *Candyland*®, *Numberland Counting Game*®, and *Chutes and Ladders*® are number games that can be bought. *Pay the Cashier* and *Count your Change* are games that help children learn money concepts. A teacher may make a shape game as follows:

Make a large cardboard triangle, a large square, and a large circle. Place these in the middle of the floor. Make small shape forms that correspond to the large triangle, square, and circle. Make at least twenty of each. To play, children draw six shapes from a bag. The teacher calls the name of a shape. If a child has that

Fig. 22-7 "Who is in the big box?"

shape, he places it on the large form. When one child uses up all his shapes, he gets to call the names. As soon as the next child uses all his shapes, he gets to call the names. The other child may draw six more shapes or stop playing.

Other basic board games were described in unit 19.

Aiming games such as bowling and drop the clothespins help children learn to count. Cardboard containers covered with Contac® paper with numerals or shape pieces attached to them make good tossing equipment to use with beanbags. Beanbags can be used in many ways. In unit 21, directions for making beanbags were given. Figure 22-8, page 184, lists ideas for using beanbags.

IDEAS FOR USING BEANBAGS

1. Balance a beanbag on your head. (Variation: Have a race.)

2. Walk a "balance beam" (2' x 4' board on some blocks) while balancing the beanbags. Walk backwards.

3. Set plastic kitchen bowls at varying distances. Try hitting the bowls.

4. Hang a balloon on a wall. Allow the child to try to hit the balloon with the beanbags. (Variation: Put several balloons on wall; assign numbers to each balloon. Hit the balloons in sequence. More advanced – add the numbers of the balloons hit.)

5. Play the "old-fashioned" game of catch, calling out the color and/or numbers of the bag being thrown. (Variation: In a group, children sit in a circle on the floor. The leader throws a bag and says one of the following: "Color, Shape, Number." The child responds, then throws to another in the group.)

6. Make a clown board to throw beanbags through, matching shapes, colors, or numbers.

Clown Throw: Purchase 4' x 4' x 1/2" Masonite at the local lumberyard (about $3.00). Using a jigsaw, cut out eyes, ears, etc., in various shapes. Paint and decorate with leftover materials. (Instead of a clown, use any animal.)

7. Put masking tape on the floor, making various shapes. Then put letters and numbers within the shapes. Throw corresponding bags into the shapes. For more advanced children, throw the beanbag on a letter and have the child give a word using that letter as the initial sound (ending sound, or middle sound). Throw two (or more) beanbags on different numbers and have the children give the sum or the difference.

8. Using a large piece of oilcloth, mark on it a large square with a smaller square inside. (A third square inside it also possible.) Throw a bag in the "big" square or the "small" square. Teach the concept of big, bigger, biggest or small, smaller, smallest. (Variation: Have two sets of squares, stand the child in one little square, throw the beanbag into the other largest square.)

9. Make a basketball hoop with an old coat hanger. Hang it inside or outside. Play basketball.

Fig. 22-8 Ways to Use Beanbags

Action games involve physical activity and mental alertness. For example, one child closes his eyes. The other children act out the following:

> I'm very, very tall,
> I'm very, very small.
> Sometimes tall,
> Sometimes small,
> Guess what I am now?

The child has to guess whether the other children are tall or small, as the last line indicates. This game can include other math ideas, also:

> I'm very very big; I'm very very thin
> I'm very very slow; I'm very very fast.

Another action game that twenty children can play requires running and catching.

> Ten little mice looking for cheese,
> They run on the floor as nice as you please.
> Ten little mice climb up high,
> They eat and eat a great big pie.
> Ten big cats out looking for fun,
> They chase the mice and make them run.

Musical Chairs is a well known action game. Children and chairs are arranged in a circle. A record player and records are available. The record is played for ten or fifteen seconds. As the music plays, children walk around the chairs. The adult removes one chair. When the music stops, the children sit. One child is left standing. This child may remove the next chair. Children can count and match: Are there more chairs, more people, or one chair for each person? Musical chairs is a math game that reinforces the concept of comparison.

Another type of game that involves a musical tool, the triangle, is called *The Striking Clock*. In this action game, children stand in a circle with their feet spread apart and rock from side to side. The teacher stands in the center with a triangle to strike the hour when the poem has been said. Children count with the striking of the triangle.

> We are swinging pendulums
> Hanging from a clock;
> As we count the hours struck,
> We rock and tick and tock.

Young children enjoy games. Board, aiming, and action games help children learn about math. Adults can observe for math skills and development as children participate.

FINGER PLAYS AND SONGS

There are many finger plays and songs that help teach children about math. The rhythm and motions that go along with finger plays and songs can be made up by the children. First, the teacher needs to say the words. Then she can sing the song. Children can repeat the words with the teacher. Next they can sing the song. Now, children can be asked how they would act out the poem or song. Appendix C gives a collection of finger plays that contain math concepts. Below are two examples of action finger poems.

> Five Little Ducks
>
> Five little ducks went out for a swim *(5 fingers swimming in a lap "pond")*
> They swam around and around and ran away *(hand behind back)*
> The Mama Duck called, "Quack, quack, quack" *(other arm held straight up from elbow, hand dropped, opens for mouth quacking).*

> But only four little ducks came back.
> Four little ducks went out for a swim, etc.
> (Repeat until "But no little ducks came back")
> Then the Mama Duck said, "Quack, quack, quack,"
> And five little ducks came back.

> Five Little Monkeys
>
> Five little monkeys, jumping on the bed, *(5 fingers, held up high)*
> One fell off and broke his head. *(Touch head)*
> We called for the doctor, and the doctor said,
> "No more monkeys jumping on the bed."
> Four little monkeys, jumping on the bed.... *(4 fingers, held up high)*
> (Continue counting backward)

Songs are also used by adults to help children explore math concepts. Three favorite math songs are *Johnny Works With One Hammer, Did You Ever See A Circle*, and *The Numeral Song*, figures 22-9, 22-10, and 22-11.

Riddles can also teach about math. Children enjoy the guessing game illustrated in figure 22-12, page 189.

Finger plays, songs, and riddles are used for many purposes. They provide a fun way to help children learn math ideas. Through action of the hands and body, rhyme, rhythm, and vocabulary are experienced by the children.

SUMMARY

Math in action means children exploring the environment. This is done through woodworking experiences and block construction. Math in action means children saying, singing, and acting out math words and ideas. Teachers incorporate these into the early childhood program to help promote math learning.

JOHNNY WORKS WITH ONE HAMMER

Other verses are as follows:

Johnny works with two hammers
Two hammers, two hammers,
Johnny works with two hammers,
Then he works with three.

Johnny works with three hammers,
Three hammers, three hammers,
Johnny works with three hammers,
Then he works with four.

Johnny works with four hammers,
Four hammers, four hammers,
Johnny works with four hammers,
Then he works with five.

Johnny works with five hammers,
Five hammers, five hammers,
Johnny works with five hammers,
Then he goes to sleep.

As each verse is sung the children do the following movements:

1. Pound on one knee with fist.

2. Pound on both knees.

3. Pound with both fists while tapping the floor with one foot.

4. Pound with both fists and tap with both feet.

5. Shake head back and forth along with all other movements.

Fig. 22-9 "JOHNNY WORKS WITH ONE HAMMER."

DID YOU EVER SEE A CIRCLE?

1. Children sit in a circle.

2. Each child is given an envelope containing 4 shapes.

3. Everyone removes the shapes from the envelope and places them on the floor in front of him.

4. Dramatize the song by holding up the appropriate shape as it is mentioned in each verse.

   "Did you ever see a circle, a circle, a circle,
   Did you ever see a circle
   Go this way and that?"

(Choose two appropriate hand movements. Up and down, side to side, circle clockwise and circle counterclockwise, forward and back, clapping hands, tapping parts of the body or the floor, and moving the entire body forward to back.)

5. Continue singing with the motions:

   "Go this way and that way
   Go this way and that way
   Did you ever see a circle
   Go this way and that?"

6. In succeeding verses, substitute square, rectangle or triangle for circle.

Fig. 22-10 "DID YOU EVER SEE A CIRCLE?"

THE NUMERAL SONG

1. The children sit in a circle.

2. The teacher keeps one set of 9-inch numerals and places one set out on the floor in the center of the circle.

3. One child is chosen to stand in the center of the circle.

4. The teacher holds up a numeral and everyone sings:

    "Oh, do you know the numeral (one)
    The numeral (one), the numeral (one)?
    Oh, do you know the numeral (one)?
    It looks just like this."

5. The child in the center picks up the matching numeral from the group on the floor and everyone sings:

    "Oh, yes, I know the numeral (one)
    The numeral (one), the numeral (one).
    Oh, yes, I know the numeral (one).
    It looks just like this."

6. Another child is chosen to go to the center and the song is repeated using another numeral. Write the numeral in the air with one hand and sing:

    "Oh, can you write the numeral (one)
    The numeral (one), the numeral (one)?
    Oh, can you write the numeral (one)?
    It looks just like this."

7. Another child is chosen to go to the center of the circle and the game is repeated.

Fig. 22-11 "THE NUMERAL SONG"

My voice is quiet
A soft tick-tick
You look at my hand
And you must be quick.

Clock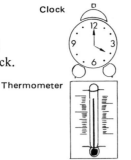

I live on a wall
With dates on my face
I tell you what day
If you're going someplace

Calendar

My color is red
When it's cold I fall
When it's hot I rise
I'm useful to all.

Thermometer

I am not a king
But my name sounds strong
I keep lines straight
And tell how long.

Ruler

Fig. 22-12 A Guessing Game

Fig. 22-13 Playing with blocks is fun!

SUGGESTED ACTIVITIES

- Visit a toy company and examine the types of blocks available for purchase. Consider how they are like or unlike those discussed in this unit.

- Examine various early education supply catalogs and check prices and types of blocks available for young children. Make a cost list of blocks you would purchase and tell why you selected them.

- Observe children playing with blocks. Determine what stage of block construction their play represents. Share your observations in class.

- Go to a local hardware and examine woodworking tools. If possible, become familiar with the use of the various pieces of equipment.

- Learn five math finger plays. Share two with your classmates.

- Share a different math action song with your classmates. This could be one that you make up.

- Plan a math action game to use with children. If possible, try it with a group of children. Share with the class what happened during this experience.

REVIEW

A.   Indicate the choice which best completes each of the following.

1. The woodworking area is

   a. Not beneficial for teaching math
   b. Well supervised
   c. Not liked by girls
   d. A piece of plywood

2. Woodworking tools should be

   a. Rusty
   b. Nails with small heads
   c. Hard wood
   d. High quality

3. One of the most basic play materials in the preschool and kindergarten is

   a. Swings   c. Blocks
   b. Guns   d. Dolls

4. Care of wood blocks includes

   a. Washing daily
   b. Allowing them to become dusty
   c. Throwing them in a pile
   d. Oiling or waxing them occasionally

5. Block accessories do not include

   a. Cars and trucks   c. Animals
   b. Paste   d. People

6. The first stage of block construction is

   a. Building vertical rows
   b. Handling blocks
   c. Making enclosures
   d. Making structures for dramatic play

7. Substitutes for building materials do not include

   a. Food   c. Hollow blocks
   b. Styrofoam forms   d. Boxes

8. An infant game well liked is

   a. Marbles

   b. Duck duck goose

   c. Musical chairs

   d. Peek-a-boo

9. Which of the following is not a board game?

   a. Candyland
   b. Ten Pin Bowling
   c. Chutes and Ladders
   d. Numberland Counting Game

10. Preschool games

    a. Should be competitive
    b. Should involve many skills
    c. Can be teacher made
    d. Should not include any action

B. Briefly answer each of the following.

1. Describe how woodworking experiences for children can be planned.

2. Describe a way to set up a block area in a school.

3. Describe how to use games for math experiences.

4. Describe how to use math finger plays and songs with children.

unit 23 math in the home

OBJECTIVES

After studying this unit, the student should be able to

• Explain the value of parents teaching children at home.

• Demonstrate the ability to implement a parent involvement approach

• List guides for parents to effectively teach their child at home

• Share with parents naturalistic, informal, and structured home math teaching experiences

His parents are the child's first teachers. The parents and the environment are most influential in the young child's development. Parents can provide the foundation for success or failure. Success is fostered by encouraging a child to be active and to explore. Failure is created by constantly offering "No" each time a child tries to do something new. Praise provides encouragement. Criticism leads to loss of interest. It is widely accepted that the child's competence — his ability to do things — is established before the age of three. The level of the child's competence is directly related to the type of parent-child interaction.

The competent child has parents who believe that the child wants to learn. They believe the child can learn. They believe learning is fun and natural! These parents recognize that the child imitates them. They recognize that even infants learn through example. They know that the more the child sees and does, the more he learns. These parents also realize that to teach the child to do things for himself is rewarding for the child.

Teachers of young children have contact with parents. Parents often ask for advice about how to help their child do things better. They want to know how their child is doing in school. They want ideas about what they should buy. They want to know how their child gets along with other children his age.

Parents are the child's most important teacher. Teachers need to emphasize this to parents. They need to reassure parents that young children are anxious and eager to learn. Teachers should help parents understand that their child's natural curiosity drives him to question, explore, and discover more about

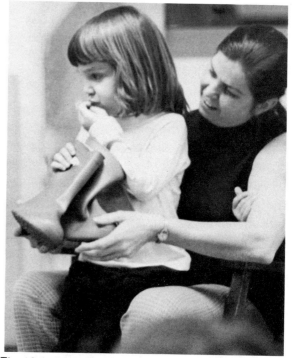

Fig. 23-1 As mother helps the young child put on her boots, the child learns.

the world in which he lives. Parents only need to provide the guidance and direction. The child looks to his parents for knowledge and help.

Parents know their child better than anyone else does. Parents can help teachers learn to understand the child better. It is important that parents and teachers communicate on a regular basis. Since parents know their child they can suggest math activities that are appropriate for the child's level of development.

APPROACHES FOR PARENT INVOLVEMENT IN MATH

There are many ways to help parents understand how to work with their children in specific skill areas. These approaches can be used for math skill development also.

One approach is to send home a newsletter each week or each month. In these newsletters, the teacher can list the skills on which the children have been working. She can tell about special experiences the children have had. A clever way to present a newsletter is to write as if the school pet is observing the children. He then informs the parent about the children. This format is often enjoyed by parents.

Some teachers like to have open house for parents. This is another approach for parent involvement. At the open house, the parents can examine the school environment and the teaching materials. Teachers can demonstrate special ways that materials are used to help children learn. Parents particularly like open houses where they can participate by working with the children's materials.

Dear Mr. and Mrs. Foster,

The children in Timmy's class have been learning how to sort things this week. They had all kinds of collections of things out in the classroom. I never thought they could organize all the piles Mrs. Olert put on the table. But guess what — the children found all the things with which to write. They made a separate pile for all the things with which to eat. They found all the items that could be used to tell time. I was really amazed at how much they liked doing this.

We visited a grocery store on Tuesday. When the children came back, Mrs. Olert had pictures of different kinds of foods. The children sorted them according to meats, vegetables, fruits, and beverages. They were excited about visiting the store.

All in all, we had a busy week. Why don't you check and see what items your child can sort for you at home?

As ever,

Freddie the Fish

Fig. 23-2 PARENT NEWSLETTER

Fig. 23-3 Grandmother shares memories of the past with this small child.

The parent-teacher conference is another parent involvement approach. This method is used to help parent and teacher learn about each child's needs. Parent concerns about how their child is doing in school can be discussed. Teachers can find out if any special home events might be affecting the child. Specific suggestions for how to best help the child handle any problems can be made. A feeling of mutual concern is needed if the child is to grow and develop to his fullest potential. This can be done by frequent parent-teacher interchanges.

GUIDELINES FOR PARENTS AS TEACHERS

If parents ask for help on how to provide learning experiences for their child, teachers need to be prepared to give them some guidelines. Parents need to be assured that natural ways of learning are best for young children. Parents need to be aware that the child's play is learning. By exploring and discovering through play, children learn. However, chil-dren do imitate what adults do. Children also model or imitate how adults feel. If the parent is positive about learning, the child will be positive, too.

As the child's first teacher, parents can provide a variety of play experiences. They can provide opportunities to repeat activities. They can provide a lot of conversation. In an atmosphere where children are free to explore and ask questions, they develop self-confidence as they learn many new skills.

There are three basic guidelines that help parents learn how to be effective teachers.

- **Patience.** The child needs a lot of time to think. If he is pressured to hurry or senses the adult wants him to work faster, he will become upset. When mistakes are made, adults should not scold the child. Learning involves mistakes. The child should be encouraged to try the activity again. Parents can suggest different ways to try the activity. If the child becomes frustrated, parents should stop the activity and try something else.

- **Repetition.** The young child needs to repeat activities. This is how he learns. He likes repetition. Familiar activities he can do well build his self-confidence. The more successful a child feels, the more he is motivated to try and learn new things.

- **Concrete experiences.** A child is interested in working with real things. He likes a variety of materials to use. The materials can be inexpensive and simple to use. The parents' imagination and the child's ideas can be used to make up concrete math experiences. This should help make these teaching sessions fun for the parent and the child.

If a parent asks the teacher for advice on how children learn math, it is important to explain about the naturalistic experiences, the infor-

Fig. 23-4 Children of all ages enjoy repeating stories such as *The Three Billy Goats Gruff*.

mal experiences, and the more structured ways of learning. The three basic guidelines just listed should be given to parents. A list of inexpensive math materials could also be shared with them. Ideas on specific math activities that can be done at home are included in this unit.

NATURALISTIC

From a very young age, children enjoy counting experiences with their parents. As mother buttons the child's sweater, she counts, "One, two, three buttons on your sweater." As she helps baby put on his shoes, she says, "One shoe, two shoes." As the family sits at the dinner table, older brother may say, "There are three boys in this house and one girl." Mother looks at the clock while carrying baby in her arms and says, "Oh dear, it's almost 2 o'clock. We will be late to the

doctor's office." Mother says, "After lunch we are going to grandmother's house. But first, we have to wash the dishes." This type of conversation goes on continually in the home. These experiences provide a natural foundation for later math learnings.

INFORMAL

As the child gets older, mother tells him to put three napkins on the table. Picture books are read to the child that contain math ideas. Birthdays are celebrated, and children count the candles. Mother makes a growth chart. The child weighs himself. Mother and child record this data on the chart. Mother measures the child's height periodically. She uses colored tape to mark changes in height. The child needs new shoes. The shoes salesperson shows the child the instrument he uses to measure feet. Mother teaches the child the proper way to dial the telephone. She teaches the child his telephone number. She helps the child learn his house number and address. As she cooks in the kitchen, she lets the child help her. As she runs errands, the child learns about the post office, the bank, the grocery store, and the gas station. Stamps have different numbers on them. Money is deposited at the bank. Fruits are weighed on a scale at the grocery store. Gas is purchased by the gallon. Math is used informally in many activities that people do daily.

STRUCTURED

Structured home math includes providing a specific place for math equipment and a specific time to work on math activities. Structured math activities in the home usually do not begin prior to age three and may be supplemental to school activities. When demonstrating math materials to the child, parents should use few words and movements. For example, the parent who is interested in teaching the child the difference between a circular shape and a square shape should first hold a cutout cardboard circular shape and show it to the child, saying, "This is a circle." The parents should then take their fingers and slowly touch the edges of the circle. Next they should hand the circle to the child and ask, "What is this?" If the child says, "I don't know," then the parent should repeat the word "circle" and put the shape away. The child's refusal indicates he is not interested or he is not ready to participate in a structured shape activity. If the child says, "Circle," then the parents say, "Show me" or "How do you know it's a circle?" The child will touch the outer edges, as he observed the parent doing. Then the activity can continue for the square shape. When the child has demonstrated that he knows a circle and a square, the parent can play games such as "Find me something square in the kitchen," or "Find

me a circle shape in the playroom." Preschool children enjoy these kinds of games at home.

When teaching structured activities, parents must be very aware of the importance of being patient. Children need time to do activities. They can be easily distracted. Structured home math activities and materials should be those in which the child is interested and which he is capable of doing. The child should choose the activities. The time spent on structured math activities may be only a few minutes each day. If parents are providing a regular school time for the child at home, they should schedule structured math within that time. The age and interest of the child should determine the length of the formal home school program. A special time may be set aside each week so that the child can anticipate that he will work on math.

HOME MATH MATERIALS

The home provides many materials that are suitable for beginning math. The button box is full of different shapes and sizes. Ribbons and yarns provide length activities. Pictures, knicknacks, and furniture can be categorized. Sets can be found in clothing, kitchen items, and father's toolbox. Folding napkins, setting the table, measuring beans, rice, or popcorn include many basic math concepts. A few commercial child math materials should be purchased for use in the home. For example, a set of blocks is a basic material. One other type of construction material such as lincoln logs or jumbo legos is also practical. Dominoes and a deck of playing cards can be used for many math activities. Balls of various sizes can be used for math and many other types of activities. Real coins can be used to make comparisons, to count, and to learn coin values. However, most math materials can be homemade. The list of activities that follows uses materials that are available in the home.

From The Kitchen	From All Around The House
egg cartons	sponges
oatmeal boxes	magazines with pictures
margarine tubs	catalogs (general, seed, etc.)
milk cartons	old crayons
milk jugs	shoe boxes
plastic lids	other cardboard boxes
egg shells	(not corrugated)
nuts	corrugated cardboard
straws	cardboard tablet backs
coffee can and lids	scraps of wallpaper, carpeting
other food cans	Contac paper
baby food jars	toothbrushes (for spattering paint)
potato chip cans	lumber scraps
orange juice cans	gift wrap (used or scraps)
yogurt, cottage cheese,	gift wrap ribbon and bows
sour cream or dip containers	old greeting cards (pictures)
milk bottle lids	newspapers
individual cereal boxes	jewelry
soft drink bottle caps	wire
plastic holder (soft drink	clothesline
6-pak carrier)	clocks
paper plates	
coffee grounds	
plastic bottles or jugs (soap, bleach, etc.)	
plastic or metal tops or lids	
styrofoam meat plates (trays)	
plastic bag twists	
cardboard rolls (paper towels, foil, etc.)	
kitchen scales	
plastic forks, spoons	

From Outside	From Sewing Scraps
rocks and stones	buttons
twigs, sticks, bark	snaps
leaves and weeds	fabric
pine cones and nuts	felt
seeds	thread
corn kernels and husks	yarn
soybeans	lace
flowers	ribbon
clay	trim
sea shells	ric rac
	spools

Fig. 23-5 MATH AIDS IN THE HOME

MATH ACTIVITIES IN THE HOME

There are many math manipulative activities that parents can do at home with children. The following activities involve counting, matching, measuring, and ordering skills. Math vocabulary can be expanded by these activities. Shape, size, and part-whole relationships can be explored. Numerals are learned, too.

─────────COUNTING─────────

Title: Clap, Bounce, Jump

Purpose: To learn how to count

Materials: Child's hands, body

Activity: Say: LET'S CLAP AND COUNT TO FOUR.   EACH TIME WE CLAP WE WILL SAY THE NUMBER.   Demonstrate.   "Clap-one" "Clap-two." Let the child practice.   Increase the range of numbers.   (Or you can say Bounce 4 times, Jump 3 times, etc.)

COUNTING/MATCHING SETS
─────────TO NUMERALS─────────

Title: Egg Container Counting

Purpose: To help the child learn how to count and sort

Materials: Egg carton with a numeral written in each cup of the carton (0-11) and the following sets: 11 paperclips, 10 buttons, 9 navy beans, 8 popcorn kernels, 7 beads, 6 sunflower seeds, 5 hairpins, 4 bottle caps, 3 safety pins, 2 poker chips, and 1 small block.

Activity: First give the child the sets with one to five items.   Say, FIND TWO POKER CHIPS AND PUT THEM WHERE IT SAYS TWO. Continue up to the numeral five.   The child will think he has been tricked when he is asked "What goes in the zero cup?"   It is important to have the zero cup, as too often it is forgotten and later confuses the child. When the child can do these well, add more cups until he can put the right set in each cup.

Fig. 23-6   Cusinaire rods have been made available to this child at home.

─────────COUNTING─────────

Title: Object Counting

Purpose: To help the child learn how to count

Materials: Containers of beans, buttons, rocks, coins, beads

Activity: Let child play with materials. Sit down with child. Say: "COUNT THE BEANS." Then see if the child can count them by himself.

─────────MATCHING─────────

Title: Find the Look Alikes

Purpose: To help child learn to match two like things

Materials: Make two sets of picture cards. One set of cards should be done drawn in one color (orange).   The second set should be the same objects but drawn in another color (green).   Put the two sets in an envelope.

Activity: Take out both sets of cards from the envelope.   Put all the orange cards in a row on the table. Say: FIND A GREEN CARD LIKE EACH OF THE ORANGE PICTURES.   PUT THEM TOGETHER. Show your child what you mean.

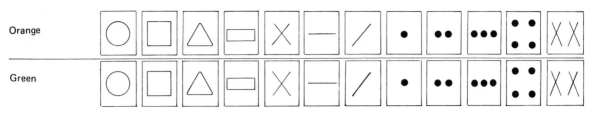

| Orange | | | | | | | | | | | | | |

Fig. 23-7 look alike cards

MEASURING

Title: Bathtub Metrics

Purpose: To help the child learn about liquid amounts using containers marked with metric vocabulary

Materials: Large, tall, thin container and one low, wide container marked with metric standards; bathtub partially filled with water.

Activity: Give the child the containers to play with during bathtime. Watch him pour from one container to the other. After he has played with the containers several times, say LET'S FILL THE TALL, THIN CONTAINER WITH .95 LITERS OF WATER. Now pour the water in the low, wide container. Ask, ARE THERE STILL .95 LITERS OF WATER? (Point to the mark.) WHY DOES THE AMOUNT OF WATER LOOK DIFFERENT?

MEASURING: WEIGHT

Title: How can it balance?

Purpose: To learn about balance and weight

Materials: Hook, flat piece of wood, heavy cord, two plastic bowls

Activity: Make the balance as shown in figure 23-9, page 200. Hang it in a spot at the child's level. Give the child several cups each half full of a different substance such as dry peas, beans, rice, etc. SAY: PUT SOME RICE IN ONE BOWL. PUT SOME RICE IN THE OTHER BOWL. IS THERE THE SAME AMOUNT IN EACH BOWL? HOW DO YOU KNOW? If they are not the same say: MAKE THEM THE SAME WEIGHT SO THE BOWLS BALANCE.

ENGLISH-METRIC EQUIVALENTS

Approximate Values

1 inch	=	25.4 mm
1 inch	=	2.54 cm
1 foot	=	0.305 m
1 yard	=	0.91 m
1 mile	=	1.61 km
1 square inch	=	6.5 cm^2
1 square foot	=	0.09 m^2
1 square yard	=	0.8 m^2
1 acre	=	0.4 hectare
1 cubic inch	=	16.4 cm^3
1 cubic foot	=	0.03 m^3
1 cubic yard	=	0.8 m^3
1 pint	=	0.47 ℓ
1 quart	=	0.95 ℓ
1 gallon	=	3.79 ℓ
1 ounce	=	28.35 g
1 pound	=	0.45 kg
1 U.S. ton	=	0.9 metric ton

SI METRIC BASE UNITS

Quantity	Unit	Symbol
Length	meter	m
Mass	kilogram	kg
Time	second	s
Temperature	kelvin	K
Electric current	ampere	A
Luminous intensity	candela	cd
Amount of substance	mole	mol
Common Unit	degree Celsius	°C

SUPPLEMENTARY UNITS

Plane angle	radian	rad
Solid angle	steradian	sr

Fig. 23-8 METRIC CHARTS

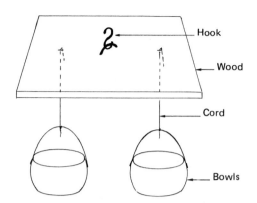

Fig. 23-9 A BALANCE TO MAKE AT HOME

Labels on figure: Hook, Wood, Cord, Bowls

————————————ORDERING————————————

Title: Measuring Cups

Purpose: To learn how to arrange items in sequence

Materials: Set of four measuring cups

Activity: Say to the child, HERE ARE FOUR MEASURING CUPS. EACH IS DIFFERENT IN SIZE. SEE IF YOU CAN ARRANGE THEM IN ORDER OF SIZE, BEGINNING WITH THE BIGGEST AND ENDING WITH THE SMALLEST.

Title: Arranging Straws

Purpose: To learn how to arrange items in sequence

Materials: Six drinking straws — cut in graduated length. The first can be one inch, the second two inches, etc.

Activity: Give the graduated straws to the child. Say: ARRANGE THE STRAWS FOR ME, ONE BESIDE THE NEXT, FROM THE SMALLEST TO THE LARGEST.

————————————SIZE/SHAPE————————————

Title: Likenesses and Differences

Purpose: To help the child notice likenesses and differences in shape

Materials: Containers: *Round* (box, bowl, plate, pie pan, cake pan, clothes basket); *Rectangular* (long cake pan, shoe box, box); *Square* (box, cake pan, clothes basket); *Objects* (magazine, ruler, sheet of paper, jar lid, ball, plates, napkin, handkerchief)

Activity: Place three containers on a table (one round, one rectangular, and one square). On the table, place many things that have the same shape as the containers. Ask the child to put into each container the things that are shaped most like the container.

Title: Size Relationships

Purpose: To help the child learn how to order small to large objects

Materials: Cutouts of dogs or any other animal the child likes (Flowers or any objects can be used.)

Activity: Make several (5 or 6) different sizes of dogs, going from small to large. Give them to the child and say: PUT THE DOGS IN ORDER ACCORDING TO SIZE. GO FROM SMALL TO LARGE.

————————————PART/WHOLE————————————

Title: Napkin Folding

Purpose: To help the child learn the concept of half through natural activities

Materials: Napkins

Activity: Give the child several napkins. Show him how to fold them in half. Then say: FOLD THE NAPKINS IN HALF AND PUT THEM ON THE TABLE. ONE FOR YOU, ONE FOR ME, ONE FOR DAD, ETC.

Title: Let's Divide

Purpose: To help your child learn parts: Half, thirds, and quarters

Materials: Pizza, apples, pears, bananas, other food

Activity: Whenever pizza, apples, pears, bananas, or other suitable foods are served, let the child help slice the food. As the parent and child slice together, talk about the pieces. Say: WE WILL FIRST CUT THE PIZZA IN HALF. NOW WE WILL QUARTER IT. NOW IT IS IN FOUR PARTS. Peel a banana — say: LET'S DIVIDE THIS INTO THREE PARTS. THREE PARTS MEAN THAT WE'LL CUT IT INTO THIRDS. Count the portions as you divide the banana.

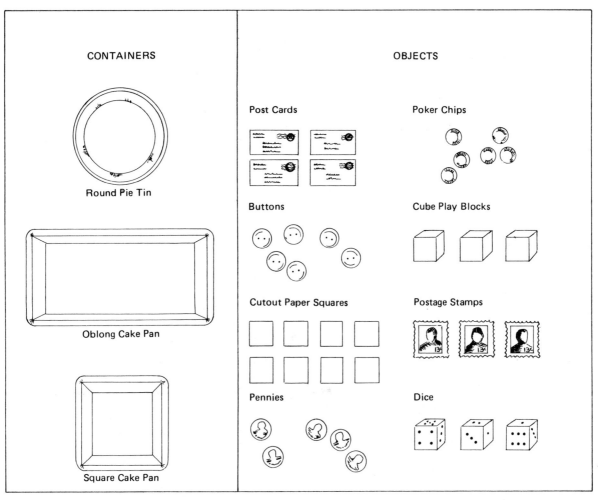

Fig. 23-10 SHAPE — LIKENESSES AND DIFFERENCES

——————SHAPES——————

Title: Find the Squares in the Room

Purpose: To learn to recognize shapes

Materials: Items of various shapes

Activity: Say to the child, FIND ALL THE SQUARE THINGS IN THE ROOM. When the child has found several, you can show him some he may have missed. If he is still interested in playing — ask him to find the round shapes in the room, etc.

Title: I'm Thinking About Everything That's Round

Purpose: To learn to recognize shapes

Materials: Natural environment

Activity: Say to the child, LET'S PLAY A GAME. FIRST, I'LL LOOK AROUND THE ROOM AND THINK OF SOMETHING THAT IS ROUND. GUESS WHAT IT IS. Once the child guesses, it is his turn to test the parent. Children love trying to trick their parents.

——————NUMERALS——————

Title: Sand Printing

Purpose: To learn how to write numerals

Materials: Colored sand (purchased at hobby stores or florist shop), aluminum foil pan

Activity: Pour sand in pan. Say, WATCH ME. Trace a numeral with finger. Have child copy.

──────────VOCABULARY──────────

Title: Tonga the Tiger

Purpose: To help the child learn about location and space

Materials: A "Tonga the Tiger" puppet (made from paper bag and crayon)

Activity: Give child verbal directions using Tonga the Tiger. Say, TONGA THE TIGER SAYS, "PUT YOUR HAND ON TOP OF YOUR HEAD." Words to use: Next to your head, at the side of your head, in front of your head, beside your head, behind your head, on your head, off your head, under your head, on top of your head, underneath your head. Other placement words to use: Inside, outside, in, out, by, beside, between, etc.

Title: Make Yourself Short

Purpose: To teach math vocabulary words

Materials: The child and parent

Activity: Say, LET'S STAND UP AND PLAY A GAME. WATCH ME. SEE IF YOU CAN DO WHAT I ASK YOU. The parent does the body actions with the child at first. Later the parent just tells the child what to do.

1. Make yourself short.
2. Make yourself shorter.
3. Make yourself tall.
4. Make yourself taller.
5. Make yourself wide.
6. Make yourself wider.

Title: What Did We Do Today?

Purpose: To learn time concepts of first, second, and last

Materials: The parent

Activity: When putting the child to bed, the parent talks about what was done first today, what happened second, what happened last. Ask the child what toy he played with *first*, etc. Ask him what TV program he watched *last*. Ask him what his *second* favorite thing was that he did today.

SUMMARY

Parents are the child's first teachers. Parents and preschool teachers need to communicate. This helps both adults know more about what the child is learning and doing. This helps adults learn about the child's needs. Approaches for parent involvement include newsletters, open houses, and conferences. To become effective teachers, parents must have patience, make use of repetition, and use concrete experiences. Naturalistic, informal, and structured experiences occur in the home through which children learn math experiences. Most home math equipment and materials are inexpensive. Most home math activities involve manipulative concrete experiences.

SUGGESTED ACTIVITIES

• Obtain approval to attend a parent and teachers conference.

• Write a parent newsletter. Share it with the class.

• Plan a mock parent workshop and share it with the class.

• Make a list of guidelines for parents to follow when teaching their children.

• Make a card file of at least fifteen home math activities.

• Volunteer to help a parent teach his child some math activity. React to what was learned about parents as teachers.

REVIEW

A. Following is an observation of a parent and child. Read it thoroughly. Then indicate what was appropriate behavior and inappropriate behavior on the part of the parent.

SITUATION: Parent is teaching child about bigger and smaller.

Parent: Deb, Let's see if you can help me find something bigger than this pencil.

Debbie: Oh, that's easy.

Parent: Well, you think you know everything.

Debbie: OK, this book is bigger than the pencil.

Parent: Yeah.

Debbie: And this brush is bigger, but the comb is the same size.

Parent: OK, that's enough, Find anything that is smaller.

Debbie: Oh, this button, key, and penny are smaller.

Parent: You sure are a smartie!

B. Answer each of the following:

1. Why do teachers of young children need to learn math activities to share with parents?

2. What are three approaches for parent involvement in math?

3. What are some guides to share with parents so they may effectively teach their child at home?

4. Describe two naturalistic, two informal, and two structured home math teaching experiences.

Appendix A

Developmental Level Assessment Tasks*

Contents

SENSORIMOTOR LEVELS

PREOPERATIONAL LEVELS

MATH WORDS

CONCRETE OPERATIONS

*Tasks which have been used as samples in the text.

SENSORIMOTOR: LEVEL 1

1A	Sensorimotor
	2 Months

1. Talk to the infant. Notice if he seems to attend and respond (by looking at you, making sounds, changing facial expression).

2. Hold a familiar object (such as a rattle) within the infant's reach. Does he reach out for it?

3. Move an object through the air across the infant's line of vision. He should follow it with his eyes.

4. Hand him a small toy. He should hold it for 2-3 seconds.

1B	Sensorimotor
	4 Months

1. Note each time you offer the infant a toy. Does he usually grab hold of it?

2. Place the infant where it is possible for him to observe his surroundings (such as in an infant seat) in a situation where there is a lot of activity. Note if his eyes follow the activity and he seems to be interested and curious.

1C	Sensorimotor
	6 Months

1. One by one, hand the infant a series of three or four nontoxic objects. Note how many of his senses he uses for exploring the objects. He should be using mouth, eyes, and hands.

2. Place yourself out of infant's line of vision. Call out to him. Note if he turns his head toward your voice.

3. When infant drops an object, note whether or not he picks it up again.

4. When he is eating, notice if he can hold his bottle in both hands himself.

5. Show the infant his favorite toy. Slowly move the toy to a hiding place. Note if infant follows with his eyes as the toy is hidden.

1D	Sensorimotor
	12 Months

1. Note if the infant will imitate the following activities as you do each one:
   a. Shake a bell (or a rattle).
   b. Play peek-a-boo by placing your open palms in front of your eyes.
   c. Put a block (or other small object) into a cup; out of a cup.

2. Partially hide a familiar toy or a cookie under a pillow or a box as the child watches. Note whether he searches for it.

3. Is he creeping, crawling and pulling himself to his feet?

4. Note whether he responds to the following verbal commands:
   a. "No, no."   b. "Give it to me."

SENSORIMOTOR: LEVEL 2

2A	Sensorimotor
	12-18 Months

1. Present him with several containers and small, safe objects. Note if he fills and dumps objects from containers.

2. Say to him, POINT TO YOUR NOSE (HEAD, EYES, FOOT, STOMACH).

3. Hide a familiar object completely. Note whether the child searches for it.

2B	Sensorimotor
	18-24 Months

1. Note if he is beginning to organize objects in rows and put similar objects together in groups during free play.

2. Ask him to point to familiar objects. POINT TO THE BALL (CHAIR, DOGGY, CAR, BLOCK, BOTTLE, DOLL, ETC.)

3. Note whether he begins to be able to name the parts of his body (two parts at eighteen months).

PREOPERATIONAL: LEVEL 3

3A*   Preoperational

Matching (unit 5): One to one   **Ages 2-3**

Play activities: Notice if he matches items one to one such as putting a small peg doll in each of several margarine containers or on top of each of several blocks which have been lined up in a row.

3B   Preoperational

Number and Counting (unit 6):   **Ages 2-3**

    "Twoness" and Rational Counting

1. When asked, HOW OLD ARE YOU? Answers, "Two" and holds up two fingers.

2. Give him two objects. HOW MANY *(name of objects)* ARE THERE? If he succeeds, try three objects. See how far he can go.

3C   Preoperational

Sets and Classifying (unit 7):   **Ages 2-3**

    Informal Sorting

As the child plays, note whether he groups toys by color, shape, size, etc. Ask him, WHICH ARE THE RED BLOCKS? WHICH CAR IS BIGGER? GET SOME SQUARE BLOCKS.

3D   Preoperational

Comparing (unit 8): Informal Measurement   **Ages 2-3**

Show the child two objects (such as two balls) of different sizes. One should be at least twice as big as the other: POINT TO THE BIG *(BALL)*. POINT TO THE LITTLE *(BALL)*. Try the same task with other concepts/words such as

large-small	heavy-light
long-short	cold-hot
fat-skinny	higher-lower

3E   Preoperational

Comparing (unit 8): Number   **Ages 2-3**

Show the child two piles of small objects (pennies, paper clips, checkers, small candies): one pile with one thing; one with six or more. Ask, WHICH HAS MORE *(object name)*? POINT TO THE ONE WITH MORE.

3F   Preoperational

Shape (unit 9): Matching   **Ages 2-3**

Show the child an object or a cutout. Ask him, FIND ONE THAT IS THE SAME. He must choose from three objects or cutouts of the same color but only one of the same shape as the first one.

3G*   Preoperational

Space (unit 10): Position   **Ages 2-3**

Have a small container (box, cup or bowl) and an object such as a coin, a checker, or a chip. PUT THE *object name* IN THE BOX (or cup or dish). Repeat using the space words (ON, OFF, OUT, IN FRONT OF, NEXT TO, UNDER).

3H*   Preoperational

Parts/Wholes (unit 11): Missing Parts   **Ages 2-3**

Have real things and/or pictures of things with parts missing. For example

Things:	A doll with a leg or arm missing
	A car with a wheel missing
	A cup with a handle broken off
	A chair with a leg gone
	A face with only one eye
	A house with no door
Pictures:	Mount picures of common things on poster board. Parts can be cut off before mounting.

Show each thing or picture and ask, LOOK CAREFULLY. WHICH PART IS MISSING FROM THIS _____?

3I	Preoperational
Ordering (unit 13): Sequence	**Ages 2-3**

Show the child three objects of the same shape but different sizes. Say, WATCH WHAT I DO. Line the objects up in order by size:

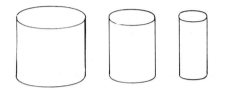

NOW I'LL MIX THEM UP (do so). PUT THEM IN A ROW LIKE I DID. If he can do the task with three objects, try it with more.

3J	Preoperational
Measuring (unit 14): Volume	**Ages 2-3**

Give the child several different size containers. Have one large container filled with small objects such as blocks or ping pong balls or filled with a substance such as beans, rice, or dry peas. Let the child experiment with pouring and filling. Look for signs that he notices that each container holds a different amount.

PREOPERATIONAL: LEVEL 4

4A	Preoperational
Matching (unit 5): Same Things/	
Related Things	**Ages 3-4**

1. Gather together four pairs of different matching objects (such as toy animals, cars, blocks, coins, etc.). Place them at random in front of the child. FIND THE ONES THAT BELONG TOGETHER. If you get no response, pick one up, FIND ONE LIKE THIS. Note if the child has a pattern of organization (such as placing the pairs side by side or in rows).

2. Collect two groups of four pairs of related objects (such as cups and saucers or cowboys and horses or flowers and flower pots). Place the objects at random in front of the child. FIND A CUP FOR EACH SAUCER (OR COWBOY FOR EACH HORSE, FLOWER FOR EACH POT). Note if the child is organized and uses a pattern for placing the objects (such as a row).

4B	Preoperational
Number and Counting (unit 6): Rote and	
Rational	**Ages 3-4**

Rote: COUNT FOR ME. START WITH ONE AND COUNT.

Rational:

1. Ask, HOW OLD ARE YOU? Does he say correct age and hold up correct number of fingers?

2. Place four objects (pennies, blocks, etc.) in front of the child. COUNT THE PENNIES (BLOCKS). HOW MANY PENNIES (BLOCKS) ARE THERE?

3. If he cannot do #2, try 2 or 3 items.

4. If he does #1 easily, try five items. If he succeeds, keep giving him larger groups. See how far he can go.

4C	Preoperational

Sets and Classifying (unit 7): Object Sorting

Ages 3-4

Place in front of the child twelve objects: 2 red, 2 blue, 2 green, 2 yellow, 2 orange, 2 purple. There should be at least five different kinds of things. For example

Color	*Object 1*	*Object 2*
red	block	car
blue	ball	cup
green	comb	car
yellow	block	bead
orange	comb	cup
purple	bead	ribbon

Give him six-ten small containers (bowls, boxes). PUT THE TOYS IN THE BOWLS (BOXES). Note the criteria (color, shape, etc.) he uses as he makes his groups.

4D   **Preoperational**

Comparing (unit 8): Informal Measurement and
                         Number   **Ages 3-4**

1. Informal measurement: See 3D

2. Number*

Place two dolls (or toy animals or cutout figures) in front of the child. I'M GOING TO GIVE EACH DOLL SOME COOKIES. (Put two cardboard cookies in front of one doll and six cardboard cookies in front of the other). SHOW ME THE DOLL THAT HAS MORE COOKIES. Go through the procedure again asking SHOW ME THE DOLL THAT HAS LESS (FEWER) COOKIES.

4E   **Preoperational**

Shape (unit 9):   Geometric Shape
                    Recognition   **Ages 3-4**

1. *With black marking pen draw a circle, a square and a triangle each on a separate card.

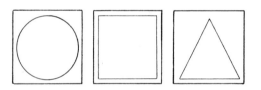

Place the cards in front of the child. POINT TO THE SQUARE. POINT TO THE CIRCLE. POINT TO THE TRIANGLE.

2. Ask the child to look around the room.

   a. FIND THINGS THAT ARE SHAPED LIKE CIRCLES.

   b. FIND THINGS THAT ARE SHAPED LIKE SQUARES.

   c. FIND THINGS THAT ARE SHAPED LIKE TRIANGLES.

4F*   **Preoperational**

Space (unit 10): Position   **Ages 3-4**

Have some small containers and several small objects. For example, four plastic glasses and four small toy animals such as a fish, a dog, a cat, and a mouse. Ask the child to name each animal (or other object) so you can use his name for each if different from yours. Line up the glasses. Place the animals so one is *in*, one *on*, one *under*, and one *between* the glasses. Then ask, TELL ME, WHERE IS THE FISH? WHERE IS THE DOG? WHERE IS THE CAT? WHERE IS

THE MOUSE? Note whether the child uses the position words in his answer.

The following tasks can be presented first between ages three and four and then repeated as the child's concepts and skills grow and expand.

4G*   **Preoperational**

Number and Counting (unit 6): Rote
                      Counting   **Ages 3-6**

COUNT FOR ME. COUNT AS FAR AS YOU CAN. If the child hesitates or looks puzzled, ask again. If he still does not count, say ONE, TWO, WHAT'S NEXT?

4H*   **Preoperational**

Number and Counting (unit 6): Rational
                      Counting   **Ages 3-6**

Place a pile of chips in front of the child (about ten for a three-year-old, twenty for a four-year-old, and thirty for a five-year-old). COUNT THESE FOR ME. HOW MANY CAN YOU COUNT?

4I*   **Preoperational**

Time and Sequence (unit 15):   Identification of a
                              Clock   **Ages 3-6**

Show the child a clock or a picture of a clock. WHAT IS THIS? WHAT DOES IT TELL US? Note whether he can name it and how much he can tell you about it: What is it for? What are the parts and what do they do? Note if he tries to tell time.

4J*   **Preoperational**

Symbols (unit 18): Recognition   **Ages 3-6**

Starting with zero show the child, one at a time, cards with the numerals from zero to ten. WHAT IS THIS? TELL ME THE NAME OF THIS.

1	2	3	4

PREOPERATIONAL: LEVEL 5

5A	Preoperational

Matching (unit 5): Same Things/
Related Things   **Ages 4-5**

Do 4A (1 and 2) using more pairs of objects.

5B	Preoperational

Number and Counting (unit 6)   **Ages 4-5**

See 4B, 4G, and 4H.

5C	Preoperational

Comparing (unit 8): Number   **Ages 4-5**

1. Present one object and a group of two objects. DOES ONE GROUP HAVE MORE? If answers yes, POINT TO THE GROUP THAT HAS MORE.

2. Use groups of up to five objects and compare as in #1. Use the terms more, less, many, and few.

5D*	Preoperational

Comparing (unit 8): Informal
Measurement   **Ages 4-5**

Present the child with two objects or pictures of objects which vary in size (height, length or width):

    FIND (POINT TO) THE BIG BLOCK.
    FIND (POINT TO) THE SMALL BLOCK.

5E	Preoperational

Shape (unit 9): Geometric Shape
Recognition   **Ages 4-5**

Present the child with separate cards each with a geometric figure drawn in black magic marker. TELL ME THE NAME OF EACH OF THESE SHAPES: triangle, circle, square, rectangle, diamond, cross.

5F*	Preoperational

Parts/Wholes (unit 11): Parts of a
Whole   **Ages 4-5**

Show the child a whole apple. HOW MANY APPLES DO I HAVE? After it is certain that the child knows there is one apple, cut the apple in half. HOW MANY APPLES DO I HAVE NOW?

5G*	Preoperational

Ordering (unit 13): Sequence/Ordinal
Number   **Ages 4-5**

Present the child with four objects or pictures of objects which vary in height, width, or all size dimensions: FIND THE (TALLEST, BIGGEST, FATTEST) or (SHORTEST, SMALLEST, THINNEST). PUT THEM ALL IN A ROW FROM TALLEST TO SHORTEST (BIGGEST TO LITTLEST, FATTEST TO THINNEST).

If the child accomplishes this task ask him, WHICH IS FIRST? WHICH IS LAST? WHICH IS SECOND? WHICH IS THIRD? WHICH IS FOURTH?

5H*	Preoperational

Time (unit 15): Language Labeling
and Sequence   **Ages 4-5**

Present the child with some pictures of daily activities such as meals, bath, nap bedtime, etc. Ask him to tell you about the pictures: TELL ME ABOUT THIS PICTURE. WHAT'S HAPPENING? Note if he uses terms such as breakfast time, lunchtime, bedtime, night, morning, etc. After he has told you about each picture ask him, PICK OUT THE PICTURE OF WHAT HAPPENS FIRST EACH DAY. After he does that, WHAT HAPPENS NEXT? Keep asking until all the pictures are lined up. Note the child's use of time words and whether the order of the pictures makes sense.

5I	Preoperational

Practical Activities (unit 17): Money   **Ages 4-5**

1. Observe the child playing store during free play using play money. Note if he demonstrates some idea of exchanging money for items and making change.

2. Show the child a nickel, dime, penny, and dollar bill. TELL ME THE NAME OF EACH OF THESE.

The following tasks can be presented first between ages four and five and then repeated as the child's concepts and skills grow and expand.

5J* Preoperational

Sets and Classifying (unit 7): Free Sort   Ages 4-5-6

Present the child with 20-25 assorted objects or pictures of objects and/or cutouts of shapes that can be grouped together by color, size, shape, or category (such as animals, people, furniture, clothing, or toys). PUT THE THINGS TOGETHER THAT BELONG TOGETHER. If the child looks puzzled, pick up an object and ask, WHAT BELONGS WITH THIS?

5K* Preoperational

Sets and Classifying (unit 7): Verbal Clue   Ages 4-5-6

Present the child with 20-25 objects or pictures of objects and/or cutouts of shapes that can be grouped together by color, shape, size, or category (such as animals, people, furniture, clothing, or toys). FIND SOMETHING THAT IS _____ . Name a specific color, shape, size, material, pattern, function, or class. Also ask, FIND SOME THINGS YOU CAN USE TOGETHER.

5L* Preoperational

Symbols (unit 18): Sequencing   Ages 4-6

Have eleven cards. On each card is one of the number symbols from zero to ten. Place them all in front of the child in random order as shown:

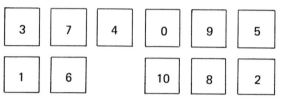

PUT THESE IN ORDER. WHICH COMES FIRST? . . .NEXT? . . .NEXT?

PREOPERATIONAL, LEVEL 6 (ENTERING KINDERGARTEN – DURING KINDERGARTEN)

6A* Preoperational

Matching (unit 5): One:one   Ages 5-6

Present the child with two groups, each with ten objects.

FIND OUT IF THERE IS THE SAME AMOUNT (NUMBER) IN EACH BUNCH (GROUP, SET). Note if he arranges each group so as to match one to one the objects from each group.

6B Preoperational

Number and Counting (unit 6): Rote and Rational Counting   Ages 5-6

1. **Rote:** COUNT FOR ME AS FAR AS YOU CAN. If child hesitates, say ONE, TWO, ... WHAT'S NEXT? (By age six he should be able to count past 10.)

2. **Rational:** Present the child with 20 objects (such as chips, coins, cube blocks, popsicle sticks, or. . .). HOW MANY _____ ARE THERE? Note the degree of accuracy and organization. (Does he place objects so as not to count any object more than once?) Note how far he can go without making a mistake. By age six he should get beyond 10.

6C Preoperational

Shape (unit 9):   Recognition and Reproduction   Ages 5-6

1. Show the child pictures of shapes to name (see 5E).

2. Reproduction
    a. DRAW A LINE
    b. COPY THE CIRCLE
    c. COPY THE SQUARE

3. *LOOK AROUND THE ROOM. FIND AS MANY SHAPES AS YOU CAN. . . .WHAT THINGS ARE SQUARE SHAPES? . . .CIRCLES? . . .RECTANGLES? . . .TRIANGLES?

6D*   **Preoperational**

Parts/Wholes (unit 11): Parts of Sets   **Ages 5-6**

Have three small child dolls (such as Weebles or Fisher-Price peg dolls). Have a box of pennies or other small objects. I WANT TO GIVE EACH CHILD SOME PENNIES. SHOW ME HOW TO DO IT SO EACH CHILD WILL GET SOME. Note how the child approaches the problem. Does he have the idea of giving one at a time to each or does he use some other method for making the set of pennies into three small sets?

6E   **Preoperational**

Ordering (unit 13): Size and Amount   **Ages 5-6**

1. Present the child with ten objects or pictures of objects which vary in size, length, height, or width as shown:

  —   ——   ———   —   ——————

———————   ————

—————   —————   —————

       FIND THE (BIGGEST, LONGEST, TALLEST, OR WIDEST) or (THE LITTLEST, SHORTEST, OR THINNEST). PUT THEM ALL IN A ROW FROM _____ TO _____ .

2. Present the child with five sets of objects consisting of 1, 2, 3, 4, and 5 objects each. Ask him to put them in sequence: PUT THESE IN ORDER FROM THE SMALLEST BUNCH (GROUP) TO THE LARGEST BUNCH (GROUP, SET).

6F   **Preoperational**

Measurement (units 14 and 15)   **Ages 5-6**

1. Weight (unit 14)

   Give the child two objects such as a plastic golf ball and a marble so that the larger object is the lighter object. Ask WHICH IS BIGGER? WHICH IS HEAVIER?

2. Time (unit 15)

   Show the child a clock. WHAT IS THIS? WHAT DOES IT TELL US?

3. Length (unit 14)

   Show the child a ruler. WHAT IS THIS? WHAT DOES IT TELL US?

6G   **Preoperational**

Practical Activities (unit 17): Money   **Ages 5-6**

1. Show the child several items or pictures of several items including money. WHERE IS THE MONEY? After he has picked out the money, ask him to name each coin or bill included. WHAT IS THE NAME OF THIS?

2. Show the child several coins. WHICH WILL BUY THE MOST? IF YOU HAVE FIVE PENNIES HOW MUCH MONEY WILL YOU GIVE ME FOR A PIECE OF CANDY THAT COSTS TWO CENTS?

Check back to 5J, 5K, and 5L, then go on to the next tasks. *The following tasks can be presented first between ages five and six and then repeated as the child's concepts and skills grow and expand.*

6H*   **Transition:Preoperational to Concrete Operations**

Ordering (unit 13): Double Seriation   **Ages 5-7**

Have two sets of ten objects or pictures of objects such that there is one item in each set that is the right size for an item in the other set. The sets could be boys and baseball bats, girls and umbrellas, chairs and tables, bowls and spoons, cars and garages, and so on. In this example hats and heads are used. The heads are placed in front of the child:

LINE THESE UP SO THAT THE SMALLEST HEAD IS FIRST AND THE BIGGEST IS LAST. Help can be given such as, FIND THE SMALLEST. OKAY, WHICH ONE COMES NEXT? AND NEXT? . . .If the child is able to line them up correctly, put out the hats:

FIND THE HAT THAT FITS EACH HEAD AND PUT IT ON THE HEAD.

6I*   Preoperational

Symbols (unit 18): One More Than   Ages 5-Older

Have eleven cards. On each card is one of the number symbols from zero to ten. Place them in front of the child in order from zero to ten. Ask, WHICH NUMERAL MEANS ONE MORE THAN TWO? WHICH NUMERAL MEANS ONE MORE THAN SEVEN? WHICH NUMERAL MEANS ONE MORE THAN FOUR? (If the child answers these then try LESS THAN.)

6J*   Preoperational/Concrete

**Sets and Symbols (unit 19):   Reproduce (Write)
                                Numerals   Ages 5-7**

Give the child a pencil, crayon, or marker and a piece of paper. WRITE AS MANY NUMBERS AS YOU CAN. Note how many the child can write and if they are in order.

6K*   Preoperational/Concrete

**Sets and Symbols (unit 19):   Matching a Set to a
                                Symbol   Ages 5-7**

Lay out in front of the child cards on which are written the numerals from zero to ten. Have a container of counters (such as chips, buttons, inch cube blocks). MAKE A SET FOR EACH NUMERAL.

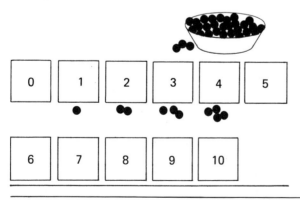

6L*   Preoperational/Concrete

**Sets and Symbols (unit 19):   Matching a Symbol
                                to a Set   Ages 5-7**

Lay out in front of the child cards on which are written the numerals from zero to ten. One at a time show the child sets of each amount. These sets can be objects or can be drawn on cards or both. PICK OUT THE NUMERAL THAT TELLS HOW MANY THINGS ARE IN THIS SET.

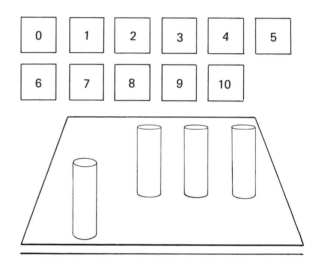

MATH WORDS: LEVEL 7

By the time the child is between 5 1/2 and 6 1/2 years of age, he should be using most of the words listed in unit 12. The following tasks can be used to find out which words the child uses in an open-ended situation. Show each picture individually.

Say, I HAVE SOME PICTURES TO SHOW YOU. HERE IS THE FIRST ONE. TELL ME ABOUT IT. For each picture, tape record or write down the child's responses. Later list all the math words. Compare this with the list of math words he uses in class.

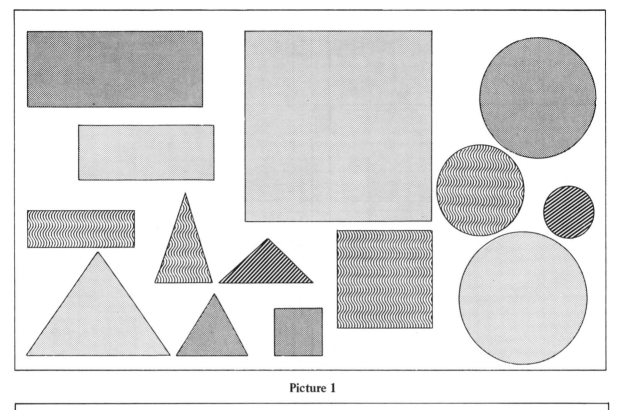

Picture 1

Picture 2

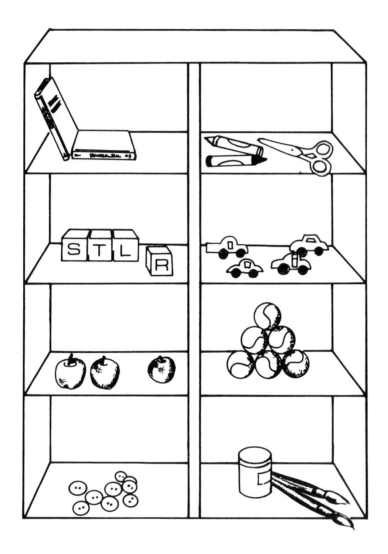

Picture 3

Picture 4

CONCRETE OPERATIONS: LEVEL 8

8A   **Concrete Operations**

Conservation of Number   **Ages 6-7**

1. **Materials**: 20 chips, blocks, or coins, all the same size, shape, and color. Give the child nine objects. You already have a row of nine. MAKE A ROW JUST LIKE THIS ONE (point to yours).

Child ☐ ☐ ☐ ☐ ☐ ☐ ☐ ☐ ☐

Adult ☐ ☐ ☐ ☐ ☐ ☐ ☐ ☐ ☐

DOES ONE ROW HAVE MORE BLOCKS (CHIPS, COINS) OR DO THEY BOTH HAVE THE SAME AMOUNT? HOW DO YOU KNOW? If child agrees to equality go on to the next tasks.

2. **Task 2**

NOW WATCH WHAT I DO. (Push yours together.)

Child ☐ ☐ ☐ ☐ ☐ ☐ ☐ ☐ ☐

Adult ☐☐☐☐☐☐☐☐☐

DOES ONE ROW HAVE MORE BLOCKS OR DO THEY BOTH HAVE THE SAME AMOUNT? WHY? (If the child says one row has more, MAKE THEM HAVE THE SAME AMOUNT AGAIN.) (If the child says they have the same amount, tell him, LINE THEM UP LIKE THEY WERE BEFORE I MOVED THEM.) Go on to task 3 and task 4 following the same steps as above.

3. **Task 3**

Child   ☐☐☐☐☐☐☐☐☐

Adult   ☐☐☐☐   ☐☐☐☐☐

4. **Task 4**

Child   ☐☐☐☐☐☐☐☐☐

Adult   ☐☐☐☐☐☐☐☐☐

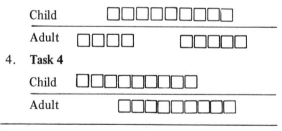

8B   **Concrete Operations**

Symbols (unit 18) and Symbols and Sets (unit 19)

                                Ages 6-7

1. Use the cards on which are written the numerals 0-9. Present the child with sets of objects (or picture cards of objects) 0-9. MATCH THE NUMBERS TO THE SETS (GROUPS, BUNCHES).

2. Have the child write the numerals 0-9 from memory. Give him pencil or crayon and paper. Say, WRITE AS MANY NUMBERS AS YOU CAN. START WITH ZERO.

8C   **Concrete Operations**

Multiple Classification (unit 20)   **Ages 6-7**

Present the child with the following group of cardboard shapes:

1. 4 squares (one each red, yellow, blue, and green)

2. 4 triangles (one each red, yellow, blue, and green)

3. 4 circles (one each red, yellow, blue, and green)

4. Make three sets of each in three sizes.

DIVIDE (SORT, PILE) THESE SHAPES INTO GROUPS, ANY WAY YOU WANT TO. After the child has sorted on one attribute (shape, color, or size) say, NOW DIVIDE (SORT, PILE) THEM ANOTHER WAY. The preoperational child will normally refuse to conceptualize another way of grouping.

8D   **Concrete Operations**

Class Inclusion (unit 20)   **Ages 6-7**

The preoperational child usually cannot think in terms of wholes and parts at the same time. Present him with twelve objects or pictures of objects such that there is one overall class and two subclasses such as:

1. twelve wooden beads of the same size and shape differing only in color (e.g. four red and 8 blue)

2. twelve pictures of flowers: eight tulips and four daisies

3. twelve pictures of animals: eight dogs and four cats

Place the objects (pictures) in front of the child in random order. PUT THE (object name) TOGETHER THAT ARE THE SAME. Then after they have grouped into two subcategories ask, ARE THERE MORE (WOODEN BEADS, FLOWERS OR ANIMALS) OR MORE (BLUE BEADS, TULIPS OR DOGS)? That is, have them compare the overall class or category with the larger subclass.

Appendix B

Children's Books With Math Ideas

Adler, Peggy. *Sets & Numbers for the Very Young.* New York: Grosset & Dunlap, 1958.

Alain. *1, 2, 3, Going to the Sea; an Adding and Subtracting Book.* Englewood Cliffs, New Jersey: Scholastic, 1969.

Allen, Robert. *Count With Me.* New York: Platt & Munk, 1968.

_____ . *Numbers.* New York: Young Scott Books, 1964.

Ambler, C. Gifford. *Ten Little Foxhounds.* New York: Grosset & Dunlap, 1958.

Asbjornsen, P.C. and J.E. Moe. *The Three Billy Goats Gruff.* New York: Harcourt Brace & World, Inc., 1957.

Atwood, Ann. *The Little Circle.* New York: Charles Scribner's Sons, 1967.

Auerbach, Marjorie. *Seven Uncles Come to Dinner.* New York: Alfred A. Knopf, 1963.

Austin, Margot. *The Three Silly Kittens.* New York: E.P. Dutton & Company, 1950.

Baker, Marybob. *Smiley Lion Book.* New York: Western Publishing Company, 1964.

Barr, Katherine. *Seven Chicks Missing.* New York: Henry L. Walch, 1962.

Barum, Arline. *One Bright Monday Morning.* New York: Random House, 1962.

Behn, Harry. *All Kinds of Time.* New York: Harcourt, Brace & World, Inc., 1959.

Beim, Jerrold. *The Smallest Boy in the Class.* New York: William R. Morrow, 1959.

Bendick, Jeanne. *All Around You.* New York: The McGraw-Hill Book Company, 1951.

Berkley, Ethel S. *Big and Little, Up and Down, Early Concepts of Size and Direction.* New York: William R. Scott, 1960.

Bishop, Claire. *Twenty-Two Bears.* New York: Viking Press, 1964.

Bishop, Claire H. and Kurt Wiese. *The Five Chinese Brothers.* Eau Claire, Wisconsin: E.M. Hale and Company, 1958.

Blair, Mary. *The Up and Down Book.* New York: Western Publishing Company, 1964.

Blegvad, Lenore. *One is for the sun.* New York: Harcourt, 1968.

Borten, Helen. *Do You See What I See?* New York: Abelard-Schuman, 1959.

Bradfield, Joan and Roger. *The Big, Happy 1-2-3.* Racine, Wisconsin: Whitman Publishing Company, 1965.

Branley, Franklyn. *Big Tracks, Little Tracks.* Crowell-Collier, 1960.

Brenner, Barbara. *Five Pennies.* New York: Alfred A. Knopf, 1963.

Bright, Robert. *My Red Umbrella.* New York: William Morrow, 1959.

Budney, Blossom and Vladimir Bobri. *A Cat Can't Count.* New York: Lothrup, Lee, and Shepard Company, 1962.

Burton, Virginia L. *Katy and the Big Snow.* Boston: Houghton-Mifflin Company, 1971.

_____ . *A Kiss is Round.* New York: Lothrup, Lee, and Shepard Company, 1954.

Campbell, Alice B., et. al. *Poems for Counting.* New York: Holt, Rinehart and Winston, Inc., 1963.

Carle, Eric. *1, 2, 3 to the Zoo.* Mountain View, California: World Publishers, 1968.

_____ . *The Ellipse.* New York: Thomas Y. Crowell, 1971.

Clure, Beth and Helen Ramsey. *Little, Big, Bigger.* California: Bromer Publishing Corporation, 1971.

Cohen, Vivian and Colette Deble. *SHAPES.* New York: WONDER BOOKS, Play School Books, 1973.

Colman, Hilda. *Watch That Watch.* New York: William R. Morrow, 1962.

Crew, Donald. *Ten Black Dots.* New York: Charles Scribner's Sons, 1968.

D'Aulasiu, Ingri. *Don't Count Your Chicks.* Garden City, New York: Doubleday & Company, 1943.

Duvoisin, Roger. *Two Lonely Ducks.* New York: Alfred A. Knopf, 1955.

Eichenberg, Fritz. *Dancing in the Moon.* New York: Harcourt, 1955.

Elkin, Benjamin. *Six Foolish Fishermen.* Chicago: Children's Press, 1964.

Emberley, Ed. *The Wing on a Flea.* Boston: Little Brown, 1961.

Everson, Dale. *Mrs. Popover Goes to the Zoo.* New York: William R. Morrow, 1955.

Federico, Helen. *Numbers.* New York: Golden Press, Inc., 1963.

_____ . *The Golden Happy Book of Numbers.* New York: Golden Press, Inc., 1963.

_____ . *1 Through 10.* Racine, Wisconsin: The Watkins-Strathmore Company, 1963.

Fehr, Howard. *Five is 5.* Little Owl Reading Time Series. New York: Holt, Rinehart and Winston, Inc., 1963.

Feltser, Eleanor B. *The Sesame Street Book of Shapes.* New York: Preschool Press, Time-Life Books, 1970.

Fisher, Margery M. *One and One.* New York: Dial Press, 1963.

Francoise. *Jeanne-Marie Counts Her Sheep.* New York: Charles Scribner's Sons, 1951.

_____ . *Springtime for Jeanne-Marie.* New York: Charles Scribner's Sons, 1951.

_____ . *What Time Is, Jeanne-Marie.* New York: Charles Scribner's Sons, 1963.

Friskey, Margaret and Katherine Evans. *Chicken Little, Count-To-Ten.* New York: Grosset & Dunlap, 1946.

_____ . *Seven Diving Ducks.* New York: David McKay Company, 1960.

Gag, Wanda. *Millions of Cats.* New York: Coward-McCann, 1945.

Grayson, Marion F. *Let's Do Finger Plays.* New York: Robert Luce, 1962.

Grender, Iris. *Playing with Numbers.* New York: Knopf/Pantheon Pinwheel Books, 1974.

Grender, Iris. *Measuring Things.* New York: Knopf/Pantheon Pinwheel Books, 1975.

Grender, Iris. *Playing with Shapes and Sizes.* New York: Knopf/Pantheon Pinwheel Books, 1975.

Gretz, Susanna. *Teddy Bears 1 to 10.* Chicago: Follet Educational Corporation, 1969.

Haley, Gail. *One, Two, Buckle My Shoe.* Garden City, New York: Doubleday & Company, 1961.

Heide, Florence & Sylvia VanClief. *How Big Am I?* Chicago: Follet Educational Corporation, 1968.

Hoban, Tana. *Count and See.* New York: The Macmillan Company, 1972.

_____ . *Push Pull Empty Full.* New York: The Macmillan Company, 1970.

_____ . *Shapes and Things.* New York: The Macmillan Company, 1970.

Hughes, Peter. *The Emperor's Oblong Pancake.* New York: Abelard-Schuman, 1962.

Ipcar, Dahlov. *Brown Cow Farm.* Garden City, New York: Doubleday & Company, 1959.

_____ . *Ten Big Farms.* New York: Alfred A. Knopf, 1958.

Jacobs, Joseph. *The Three Wishes.* Whittlesly Howe, 1961.

Jerichin, Cecile. *Hello! Do You Know My Name?* New York: G.P. Putnam's Sons, 1963.

Kaufmann, Joseph. *Big & Little.* New York: Golden Press, Inc., 1966.

Kay, Helen. *One Mitten Lewis.* New York: Lothrup, Lee, and Shepard Company, 1955.

Kent, Jack. *Just Only John.* New York: Young Readers Press, 1968.

Kessler, Ethel and Leonard. *Are You Square?* Garden City, New York: Doubleday & Company, 1966.

Kessler, Leonard. *I Made a Line.* New York: Grosset & Dunlap, 1962.

_____ . *The Worm, The Bird and You.* New York: Dodd, Mead & Company, 1962.

Keyes, Juliet. *Two-Little Birds and Three.* Boston: Houghton-Mifflin Company, 1960.

Kirn, Ann. *Nine In a Line.* New York: Norton, 1966.

Klein, Leonore. *Just A Minute: A Book About Time.* New York: Harvey House, 1969.

_____ . *What Is An Inch?* New York: Harvey House, 1966.

Kohn, Bernice. *Everything Has a Size.* Englewood Cliffs, New Jersey: Prentice-Hall, Inc., 1964.

Kohn, Bernice and Aliki Brandenberg. *Everything Has a Shape.* Englewood Cliffs, New Jersey: Prentice-Hall, 1964.

Krahn, Fernando and Maria De LaLuz. *The Life of Numbers.* New York: Simon and Schuster, 1970.

Krasilorsky, Phyliss. *The Very Little Girl.* Garden City, New York: Doubleday & Company, 1958.

_____ . *The Very Little Boy.* Garden City, New York: Doubleday & Company, 1962.

Krauss, Ruth. *The Growing Story.* New York: Harper Brothers, 1947.

Kruss, James. *3 x 3: Three by Three.* New York: The Macmillan Company, 1965.

Kuskin, Karla. *James & the Rain.* New York: Holt, Rinehart and Winston, Inc., 1963.

_____ . *Square As A House.* New York: Harper & Row, 1960.

Langstaff, John. *Over in the Meadow.* New York: Harcourt, Brace & World, Inc., 1957.

Lathan, Jean L. and Bee Lewi. *The Cuckoo That Couldn't Count.* New York: The Macmillan Company, 1961.

Le Sieg, Theo. *Ten Apples Up on Top.* New York: Random House, 1961.

Leaf, Munro. *Metric Can Be Fun.* New York: J.B. Lippincott, 1976.

Lerner, Sharon. *Square is a Shape.* Minneapolis: Lerner, 1970.

Lionni, Leo. *Inch by Inch.* New York: Ivan Oblensky, Inc., 1960.

Llodhas, Sorche Mic. *All In The Morning Early.* New York: Holt, Rinehart and Winston, Inc., 1963.

Marino, Dorothy. *Edward and the Boras.* New York: J.B. Lippincott, 1957.

Martin, Patricia Miles. *The Cat! 1-2-3.* New York: Putnam, 1969.

Matthiesen, Thomas. *Thing to See: A Child's World of Familiar Objects.* New York: Platt & Munk, 1966.

Mayres, Patrick. *Just One More Block.* Chicago: Albert Whitman and Company, 1970.

McLeod, Enile. *One Snail and Me.* Boston: Little Brown, 1961.

_____ . *The Seven Remarkable Bears.* Boston: Houghton-Mifflin Company, 1954.

Memling, Carl. *I Can Count.* New York: Golden Press, Inc., 1973.

Moore, Lillian. *My Big Golden Counting Book.* New York: Golden Press, Inc., 1957.

Otto, Margaret. *Three Little Dachshunds.* Little Owl Reading Time Series; New York: Holt, Rinehart, and Winston, Inc., 1963.

Oxenbury, Helen. *Numbers of Things.* New York: Franklin Watts, Inc., 1968.

Peppe, Rodney. *Circus Numbers.* New York: Delacorte, 1969.

Presland, John. *How Many?* Wilts, England: Child's Play (International) Ltd., 1975.

Raabe, Janis Asad. *Six Kids.* Cleveland: Modern Curriculum Press, Inc., 1974.

Ramirez, Carolyn. *Small as a Raisin, Big as the World.* Irvington-on-Hudson, New York: Harvey House, 1961.

Rand, Ann and Paul. *Little I.* New York: Harcourt, Brace & World, Inc., 1962.

Reiss, John J. *Numbers, a book.* Scarsdale, New York: Bradbury, 1971.

Rolf, Myller. *How Big is a Foot.* Patterson, New Jersey: Atheneum Press, 1962.

Rutland, Jonathan. *Time.* New York, New York: Grosset & Dunlap, 1976.

Scarry, Richard. *Best Counting Book Ever 1-2-3.* New York: Children's Television Workshop, Random House, 1973.

Schlein, Mirian. *Heavy Is a Hippopotamus.* New York: Scott, 1954.

_____ . *Shapes.* Eau Claire, WI: E.M. Hale and Company, 1952.

Schott, Andrew F. *The Littlest One.* Tools for Education, Inc., 1961.

Scott, Binder and J.S. Thompson. *Rhymes for Fingers and Flannel Boards.* New York: Webster Publishing Company, 1960.

Sendak, Maurice. *One Way Johnny.* New York: Harper & Row, 1962.

Seuss, Dr. *Dr. McElligot's Pool.* New York: Random House, 1949.

_____ . *One Fish, Two Fish, Red Fish, Blue Fish.* New York: Random House, 1960.

Shapp, Charles and Martha. *Let's Find Out What's Big and What's Small.* New York: Franklin Watts, Inc., 1959.

_____ . *Let's Find Out What's Light and What's Heavy.* New York: Franklin Watts, Inc., 1960.

Shapur, Freda. *Round and Round and Square.* New York: Abelard-Schuman, 1965.

Shuman, Diane. *My Counting Book.* Chicago: Rand-McNally, Inc., 1963.

Slobodkin, Louis. *Millions and Millions.* New York: Vanguard Press, Inc., 1955.

Slobodkina, Esphyr. *The Clock.* Eau Claire, Wisconsin: E.M. Hale and Company, 1961.

Stanek, Muriel. *One, Two, Three for Fun.* Chicago: Albert A. Whitman, 1967.

Steiner, Charlotte. *Ten in a Family.* New York: Alfred A. Knopf, 1960.

Stobbs, William, Illustrator. *The Story of the Three Bears.* New York: The McGraw-Hill Book Company, 1965.

Sullivan, Joan. *Round is a Pancake.* New York: Holt, Rinehart, and Winston, 1963.

Todd, Mary Fidelis. *A B C and 1 2 3.* Whittlesly Howe, 1955.

True, Louise. *Number Men.* Chicago: Children's Press, 1962.

Tudor, Tasha. *1 is One.* New York: H.Z. Walch, 1956.

Ungerer, Tomi. *One, Two, Where's My Shoe?* New York: Harper & Row, 1964.

_____ . *Snail, Where Are You?* New York: Harper & Row, 1962.

Vogel, Ilse-Margaret. *1 is No Fun but 20 is Plenty!* Eau Claire, Wisconsin: Hale, 1965.

Watson, Nancy D. *Annie's Spending Spree.* New York: Viking Press, 1957.

_____ . *What is One.* New York: Alfred A. Knopf, 1954.

Webber, Irma. *Up Above and Down Below.* New York: William R. Scott, 1958.

Wallu, Leslie. *A Book to Begin on: Numbers.* New York: Holt, Rinehart, and Winston, 1959.

_____ . *A Book to Begin on: Time.* New York: Holt, Rinehart, and Winston, 1959.

Welsh, Margaret Wells. *How Long Is It?* A Learn about Time Activity Book. Racine, Wisconsin: Golden Press; Western Publishing Company, 1964.

Wildsmith, Brian. *Brian Wildsmith's 1, 2, 3's.* New York: Franklin Watts, Inc., 1955.

Wilkin, Eloise. *So Big.* New York: Western Publishing Company, 1968.

Wing, Henry R. *Ten Pennies for Candy.* Little Owl Reading Time Series; New York: Holt, Rinehart and Winston, 1963.

_____ . *What is Big?* Little Owl Reading Time Series; New York: Holt, Rinehart and Winston, 1963.

Wooley, Catherine. *Two Hundred Pennies.* New York: Morrow, 1958.

Yolen, Jane H. *See This Little Line?* New York: David McKay Company, 1963.

Ziner, Feenie. *Counting Carnival.* New York: Coward-McCann, 1962.

Zolotow, Charlotte. *One Step, Two.* New York: Lothrup, Lee and Shepard Company, 1955.

_____ . *Over and Over.* New York: Harper & Row, 1957.

Zolotow, Charlotte. *The Sky Was Blue.* New York: Harper & Row, 1963.

The Sesame Street, 1, 2, 3 Story Book. New York: Children's Television Workshop, Random House, 1973.

One, Two, Buckle My Shoe; a Book of Counting Rhymes. Garden City, New York: Doubleday & Company, 1964.

Appendix C

FINGER PLAYS

The Monkey

One little monkey was looking at you,
He was joined by another and then there were two.
Two little monkeys playing in a tree,
Were joined by another and then there were three.
Three little monkeys saw one more,
She came to play with them,
And then there were four.
Four little monkeys happy to be alive,
Were joined by another,
And then there were five.

Two Apples

Way up high in the apple tree,
Two little apples smiled at me.
I shook the tree as hard as I could.
Down fell the apples —
m-m-m-m-m-m! Were they good!

1, 2, 3 Clap

Your hand on your shoulder,
And then on your toes.
Your right hand on your knee,
Then on your lap it goes.
Place the left hand on your nose.
Touch your left hand to the floor.
Clap your hands 1, 2, 3
Would you like to do that some more?

1, 2, 3 Clap

See the birds in the tree,
Now let's count them: 1, 2, 3
Now do it backwards: 3, 2, 1
Away fly the birds, every one.

Feet and Hands

Two little feet go jump, jump, jump,
Two little hands go thump, thump, thump.
One little body turns round and round.
One little child sits quietly down.
Two little feet go tap, tap, tap.
Two little hands go clap, clap, clap.
A quick little leap up from the chair.
Two little arms reach high in the air.

Two Little Blackbirds

Two little blackbirds standing on a hill.
This one is Jack, and this one is Jill.
Fly away Jack; Fly away Jill.
Come back, Jack; come back, Jill.
Now there are two little blackbirds
standing on the hill.

I See Three

I see three: one, two, three
Three little bunnies reading the funnies.
I see three: one, two, three
Three little kittens, all wearing mittens.
I see three: one, two, three
Three little bears climbing the stairs.
I see three: one, two, three
Three little ducks riding in trucks.
I see three: one, two, three
Three little frogs sitting on logs.
I see three: one, two, three
Three little bees, buzzing in trees.
(Children may suggest more rhyming verses.)

Four Little Snowmen

I built four snowmen
All in a line.
They looked so strong,
They looked so fine.
Along came a child
As silly as can be.
Who packed down one snowman,
And that left three.
Down came the rain
From a sky of blue,
Smush went one snowman,
And that left two.
The clouds passed by
And out came the sun.
It melted one snowman,
And that left one.
One little snowman
Alone once more,
I built three more again,
And that made four.

Five Soldiers

Here are five soldiers standing by the door,
One marches away and then there are four.
Four little soldiers looking at me,
One falls down and then there are three.
Three little soldiers going to the zoo,
One goes home and then there are two.
Two little soldiers left all alone,
One says, "Good-bye" and then there is one.
One little soldier lifts up his gun,
He marches away and then there are none.

Here is Thumbkin

Here is thumbkin, Number 1
He is big, but see him run.
Here is pointer, Number 2
He can show the way for you.
This is long man, Number 3
He's the tallest one you see.
This is lazy, Number 4
Try to lift him just once more.
Here's the baby, Number 5
Tiny, yes, but quite alive.
Now move your fingers, do a dance!
Lift them high, like horses prance.
Now into their houses they all creep,
Make a fist — they've gone to sleep!

Five Little Kittens

Five little kittens sitting on the floor,
One ran away and then there were four.
Four little kittens playing round a tree,
One went to sleep and then there were three.
Three little kittens beginning to "Mew,"
One climbed the tree and then there were two.
Two little kittens playing in the sun,
One went home and then there was one.
One little kitten left all alone.
He chased a mouse and then there was none.

Five Little Chickadees

Five little chickadees sitting by the door (5 fingers)
One flew away and then there were four.
(Chorus)   Chickadee, chickadee, happy and gay
           Chickadee, chickadee, fly away.
Four little chickadees sitting in a tree
One flew away and then there were three.
(Chorus)
Three little chickadees sitting just like you,
One flew away and then there were two.
(Chorus)
Two little chickadees having lots of fun,
One flew away and then there was one.
(Chorus)
One little chickadee sitting all alone.
It flew away and there were none.
(Chorus)

Here is The Beehive

Here is the Beehive.
Where are the Bees?
Hidden away where nobody sees.
Soon they come creeping out of the hive:
One, Two, Three, Four, Five.

Five Little Ducks

Five little ducks standing in a row,
They walk to the water, going to and fro.
This duck cries, "Quack, quack, quack."
This one says, "Let's go back."
This one yells, "Stay in line."
This one says, "The water's fine."
The last little duck just pecked at the ground,
Eating the bugs that crawled around.

Birthday Years

This is _____ .
It's his birthday today.
He's _____ years old, so they say.
Let's count his age as we clap our hands,
To let him know we think he's grand.
1, 2, 3, 4, 5.

Add One More

This is one cat,
Now two cats I see.
Another cat will make it three.
I add another,
Then there are four.
Mother cat joins them.
To make one more.
Now let's count them: 1, 2, 3, 4, 5.

Addition

One and one are two. That I always knew.
Two and two are four. They could be no more.
Three and three are six. Whether stones or sticks.
Four and four are eight, if I can keep them straight.
Five and five are ten. Let's try it all over again.

Ten Fingers

I have ten little fingers.
And they all belong to me.
I can make them do things.
Would you like to see?
I can shut them up tight
Or open them wide.
I can put them together
Or make them all hide.
I can make them jump high,
I can make them jump low,
I can fold them up quietly,
And fold them just so.

USE OF ORDINAL NUMBERS

Little Dogs

The first little dog barked very loud.
The second little dog ran after a crowd.
The third little dog said, "Let's eat, let's eat!"
The fourth little dog said, "Let's have meat!"
The fifth little dog said, "I think I will stay,
in my own backyard and sleep all day!"

Five Little Snowmen

Five little men all made of snow,
Five little snowmen, all in a row,
Out came the sun and stayed all day.
The first little snowman melted away.
(repeat with second, third, fourth, and fifth.)

Five Little Pumpkins

Five little pumpkins sitting on a gate
The first one said, "Oh, my, it's getting late."
The second one said, "There are witches in the air."
The third one said, "But we don't care."
The fourth one said, "Let's run, and run, and run."
The fifth one said, "I'm ready for some fun."
"Ooo" went the wind, and out went the light.
And the five little pumpkins rolled out of sight.

Ten Little Snowmen

Ten little snowmen standing in a line;
The first one melted and then there were nine.
Nine little snowmen standing tall and straight;
The second one melted and then there were eight.
Eight little snowmen, white as in heaven;
The third one melted and then there were seven.
Seven little snowmen with arms made of sticks;
The fourth one melted and then there were six.
Six little snowmen looking alive;
The fifth one melted and then there were five.
Five little snowmen with mittens from the store;
The sixth one melted and then there were four.
Four little snowmen beneath a green pine tree;
The seventh one melted and then there were three.
Three little snowmen with pipes and mufflers, too;
The eighth one melted and then there were two.
Two little snowmen standing in the sun;
The ninth one melted and then there was one.
One little snowman standing all alone;
He started to run, run, run, and then there was none.

POEMS TO TEACH SIZE AND SHAPE

Big and Fat

Santa Claus is big and fat.
He wears black boots
and a bright red hat.
His nose is red,
just like a rose,
And he Ho-Ho-Ho's
From his head to his toes.

Big Round Cookie

I looked in the cookie jar, and what did I see?
A big round cookie Mother put there for me.
Mother looked in the cookie jar,
But she didn't see
The big round cookie she put there for me.

The Ball

A little ball,
A larger ball,
The largest ball I see.
Now let's count them:
One, Two, Three.

Fingers

My fingers make a circle,
One that's very small.
Now I make a big one,
Just like a rubber ball.
This is my ring finger,
This one is the smallest.
I put all my fingers together,
To see which is the tallest.

Bibliography

Baratta-Lorton, M. *Work Jobs.* Menlo Park, California: Addison-Wesley Publishing Company, 1972.

Baratta-Lorton, M. *Work Jobs for Parents.* Menlo Park, California: Addison-Wesley Publishing Company, 1975.

Behrman, P. and Millman, J. *How Many Spoons Make A Family.* San Rafael, California: Academic Therapy Publications, 1971.

Biggs, E. *Mathematics for Younger Children.* New York: Citation Press, 1971.

Croft, D. and Hess, R. *An Activities Handbook for Teachers of Young Children.* Boston: Houghton Mifflin Company, 1975.

Greenburg, P. *Bridge to Reading, Section VII. Free Play Enrichment Projects (Music and Movement, Mathematics, and Cognitive Skills and Social Studies).* New Jersey: General Learning Corporation, 1973.

Holt, M. and Dienes, Z. *Let's Play Math.* New York: Walker and Co., 1973.

Kamii, C. and Devries, R. *Piaget, Children and Number.* Washington, D.C.: National Association for the Education of Young Children, 1976.

Mathematical Skills and Scientific Inquiry. Threshold Early Learning Library, Volume 2. New York: The Macmillan Company, 1970.

Nuffield Mathematics Project.* *Beginnings.* New York: John Wiley and Sons, 1967.

Nuffield Mathematics Project. *Mathematics Begins.* New York: John Wiley and Sons, 1967.

Nuffield Mathematics Project. *I Do and I Understand.* New York: John Wiley and Sons, 1967.

Nuffield Mathematics Project. *Pictorial Representations.* New York: John Wiley and Sons, 1967.

Nuffield Mathematics Project. *Environmental Geometry.* New York: John Wiley and Sons, 1967.

Payne, J.N. *Mathematics Learning in Early Childhood.* Thirty-seventh Yearbook, National Council of Teachers of Mathematics, Reston, Virginia, 1975.

Sharp, E. *Thinking is Child's Play.* New York: Avon Books, 1969.

Throop, Sara. *Mathematics for the Young Child.* Belmont, California: Fearon Publishers Inc., 1974.

*Some of these Nuffield books are out of print but if available in a library or elsewhere they are excellent supplementary resources.

Answers to Review Questions

SECTION 1 MATH DEVELOPMENT IN YOUNG CHILDREN

Unit 1 How Math Develops

A. *Math development* refers to the changes in math ideas and skills that take place due to growth and experience over time.

B. 1. d   2. a   3. e   4. c   5. b

C. 1. CO   2. SM   3. P

D. Mary is a conserver. She is not fooled by the physical change in the appearance of the clay. It is important to know that a child can conserve before he is taught formal mathematics such as addition and subtraction.

Unit 2 How Math Is Learned

A. *Naturalistic*: child initiated; adult provides environment, observes and gives attention when needed.
Informal: adult steps in and takes advantage of a learning opportunity; not preplanned.
Structured: adult plans and initiates; may be at a specific time or when an opportunity presents itself.

B. 1. naturalistic   4. informal
     2. naturalistic   5. naturalistic
     3. structured

C. 1. Good chance to observe a naturalistic activity. Make a note that Richard and Diana can match one to one for table setting. (See unit 5)

     2. Plan a structured series of lessons which will help the children tell the difference between squares and rectangles. (See unit 9)

     3. Teach informally. "How can we find out, boys?" Encourage them to match one to one and/or count to find out who has more crayons. (See unit 6)

     4. Teach informally if Pete is unable to solve the problem on his own. "What's wrong, Pete? Check the the shape of the peg and hole. Are they the same? Look closely." Give him plenty of time to solve the problem on his own first. (See unit 9)

     5. Plan some structured and informal experiences which will teach time words. Informally use the words often.

Unit 3 How Math Is Taught

A. 1-5;   2-6;   3-2;   4-3;   5-4;   6-1

B. 1. Assess   3. Choose objectives   5. Teach
     2. Evaluate   4. Plan experiences   6. Select materials

C. 1. Select materials   3. Evaluate   5. Teach
     2. Choose objectives   4. Assess   6. Plan experiences

D. The teacher is able to plan experiences for the child which start from his present stage. There is no need for guessing. The child is able to succeed and the teacher sees positive results.

E. If the child has reached his objective, he can go on to the next one. If not, the teacher can reassess the objectives, the activities, and the materials and decide whether the child needs more of the same or whether she needs to question him further and choose a different objective.

F. Miss Collins does not use the six-step approach. By assuming that all children are the same she probably causes both the slower and the advanced children many frustrations. It is not safe to assume that the same program is right for every child or for every group from year to year.

Unit 4 How to Assess the Child's Developmental Level

A. Find out the child's age. Start with assessment questions from that level in your assessment task kit. If the child has difficulty, go to a lower level. If the questions are answered easily, go to a higher level. Finding out where the child is helps the teacher find the right objectives for the child, plan lessons, and choose materials that have a chance of really working. This way the child has less chance of being bored with activities that are too easy or frustrated by activities that are too hard.

B. 1. There is too much going on which takes the child's mind off of the task. Try to find another place or see if the staff might be willing not to use the lounge for an hour or so on the days when she needs to do assessments.

2. Before doing an assessment, the person should get to know all the children so they will feel comfortable with her.

3. Mr. Flores must learn to give the child support through nods, smiles, words, and gestures.

4. Mrs. Raymond must remember not to use words like right and wrong but to accept whatever the child says or does.

C. 1. level six;   3. level two;   5. level four;
    2. level eight;   4. level two;   6. level one

D. 1. Explain the importance of assessing sensorimotor learning and the basic value of sensorimotor learning to later success.

2. Start questioning at level five before choosing objectives.

3. Tim seems normal for his age. Try some level four tasks before choosing objectives.

4. Level five is hard for Mary. She will need some structured counting lessons.

5. Bobby seems to be a well developed five-year-old. He could be given some higher level tasks.

SECTION 2 THE BASICS OF MATH

Unit 5 Matching

A. Matching (or one to one correspondence) is the young child's way of finding out if one group has the same amount as another. For example, he wants to buy five gum balls for one penny each. The clerk puts the five gum balls on the counter and the child puts a penny next to each one.

B. 1. Structured   3. Naturalistic   5. Structured
    2. Naturalistic   4. Informal

C. Vary perceptual difficulty, concreteness, whether or not the pairs are joined, whether the groups have the same number, and the number of things to be matched. Perceptually, groups of things that are different are easiest to match (such as spoons and bowls). Physically joining the pairs makes it easier to tell if there is one for one. Real things are easier to match than picture patterns. Larger groups are harder than small groups. If one group has more, it is harder for the child to check his solution.

D. 1. b   2. b   3. a   4. b   5. a

Unit 6 Number and Counting

A. 1. Rote counting involves saying number names in order from memory.

   2. Rational counting involves attaching each name in order to a series of objects in a group.

B. Correct statements are 2, 3, 4, 6.

C. Rote counting: 2, 3, 6

   Rational counting: 1, 4, 5

D. 1. Give praise and attention, such as saying, "Good for you, Randy."

   2. Give praise and attention, such as "Right, Tanya. We both have two eyes."

   3. This should be easy for a five-year-old. Plan some activities for individual help.

   4. Find out if Bobby can rote count correctly. Give him help with rote counting first. Then work on rational counting starting with groups of two or three.

   5. Ask, "How many children are at this table?" "How many people?" "How many desserts do we need today?"

   6. These children need some help. Plan some structured lessons for them.

Unit 7 Sets and Classifying

A. A *set* is a name for one or more things which are placed together in one group on the basis of a common feature. A set with nothing in it is called an empty set.

B. The following steps could be followed:

   1. Show the children groups of things they have played with such as blocks, toy cars, and people figures.

   2. Put each group of objects together and label them:

   THIS IS A SET OF BLOCKS.
   THIS IS A SET OF CARS.
   THIS IS A SET OF PEOPLE.

   3. Ask the children (pointing to a set), WHAT IS THIS? YES, IT IS A SET OF _____ .

   4. Give more examples.

   5. Have the children search around the room for sets.

   6. Have the children think of sets they have at home.

C. *Classification* is any activity that involves sorting and grouping.

D. The features are:

1. f	3. i	5. g	7. d	9. k	11. e
2. h	4. j	6. a	8. b	10. c	

Unit 8 Comparing

A. 1. Comparing takes place when the child finds a relationship between two things or sets of things on the basis of an attribute. "Mom, this cup is *bigger* than this other cup."

   2. Informal measurement includes size, length, height, weight, and speed. "Look, I'm taller than you." says Jimmie to Tammie.

   3. Number measurement involves looking at two sets and deciding if they have the same number of things or if one has more things. "I have more blocks than you have."

B. 1. d   2. b   3. f   4. a   5. c   6. e

C. 1. Matching, counting, and classifying.

2. Give the child a big box and a little box. Say, POINT TO THE BIG BOX. POINT TO THE SMALL BOX. Place two cutout clowns and ten cutout balloons on the table. Put two balloons by one clown and eight balloons by the other. Say, SHOW ME THE CLOWN THAT HAS MORE BALLOONS. WHICH CLOWN HAS FEWER BALLOONS?

3. (a) In conversation;   (b) using materials.

4. (a) Situations that involve child physically;   (b) Labeling attributes of objects.

5. (a) The teacher plays the xylophone. She asks the child which tones are loud/soft; which tones are higher/lower.

    (b) The teacher has a box of weights. She asks which are heavier/lighter.

Unit 9 Shape

A. 1. Pillow: Naturalistic

2. Round Waffle: Informal

3. Washing: Naturalistic

4. Square plate: Informal

5. Ride to school: Naturalistic

6. Block building: Naturalistic

7. Square building: Informal

8. Pipe cleaner art: Naturalistic

9. Circle people: Informal

10. Snack: Naturalistic/Informal

11. Story time: Structured

12. Musical shape game: Structured

13. Attribute block play: Informal

B. 1. By observation.

2. Tell the child to look around the room and find square shapes, circles, rectangles, and triangles.

3. Take a feeling bag. Put in different shapes. Have the child use his sense of touch to label and discriminate shapes.

4. *Discrimination* assesses whether the child can see that one form has a different shape from another form.

Labeling assesses whether the child can find a shape when the name is given and whether he can name a shape when a picture is shown to him.

Matching assesses whether the child can find a shape like one shown to him.

Sorting assesses whether the child can separate a mixed bunch of shapes into sets.

C. 1. b   2. a   3. b   4. c   5. d   6. c

Unit 10 Space

A. 1. Note the child's use of space words.
Note the child's use of organization and pattern arrangement during free play.
Note the child's use of construction materials.
Note the child's use of his own body in space.

2. Teacher uses space words.
Teacher gives space directions.
Teacher asks space questions.
Teacher provides experiences in which children manipulate materials.

3. Teacher sets up obstacle course.
Teacher and children play "Find Your Friend" game.
Teacher sets up activity where children have to place objects in various spatial relationships.

B. 1. c   3. d   5. a, b, c   7. d   9. a   11. d   13. a   15. b
2. e   4. e   6. c   8. b   10. b   12. d, e   14. e

Unit 11 Parts and Wholes

A. 1. c   3. b   5. c   7. a   9. c

    2. a   4. c   6. b   8. b   10. a

B. 1. Through exploring objects and adult-child verbal interaction.

    2. a. Things, people and animals have parts.
        b. Sets can be divided into parts.
        c. Whole things can be divided into smaller parts or pieces.

    3. a. Teacher uses the words such as part, whole, divide, and half.
        b. Teacher gives the child tasks that involve part/whole relationships such as passing out the crackers.

    4. a. Teacher shows the child a broken toy and asks child to tell what part is missing.
        b. Teacher has two or more containers and small objects which the child has to divide into smaller groups.

    5. Teacher shows child sets of picture cards that contain pictures of things with missing parts. Child is asked which part is missing.

    6. Teacher observes child's use of part/whole words and skills in his daily activities.

Unit 12 The Language of Math

A. 1. by what the child does
        by what the child says

    2. pair, sets, pattern, big-small, circle, square, off, on top of, divide, pieces.

    3. fewer, fewest; more, most; medium, quart, gallon, ounces, liter, meter

B.

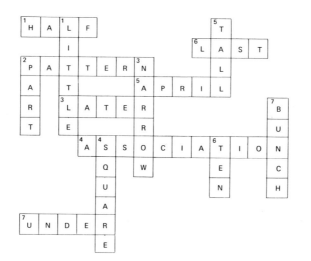

SECTION 3 USING THE BASICS

Unit 13 Ordering

A. These items apply to ordering: 2, 4, 5, 6, 7, 9.

B. 1. Ordering is the act of comparing more than two things.

    2. Another name for ordering is seriation.

3. The four basic types of ordering activities are

    a. putting things in sequence by size.
    b. making a one to one match between two sets of related things.
    c. placing sets of different things in order from the least to the most.
    d. learning ordinal numbers such as first, second, third.

C.   1. c   3. a   5. e   7. f   9. d

     2. d   4. a   6. b   8. a/d   10. a

Unit 14 Measurement: Volume, Weight, Length, and Temperature

A.   1. c   2. e   3. b   4. a   5. d

B.   The measurement stages are

     1. A *play* stage in which the child imitates other children or adults, pours objects from one container to another, lifts things.

     2. A *comparison* stage in which the child checks to see what is heavier or lighter, longer or shorter, and so on.

     3. An *arbitrary unit* stage in which the child uses anything as a unit of measure.

     4. An understanding of the need for standard units.

     5. The child begins to *use* and understand the standard units of measurement.

C.   1. Stage 2   2. Stage 5   3. Stage 1   4. Stage 4   5. Stage 3

D.   Observe the child and note how he imitates adult use of measurement. Also observe whether the child solves problems by using informal measurements such as comparisons.

Unit 15 Measurement: Time

A.   *Sequence of time* has to do with the order of events.

B.   Duration of time has to do with how long an event takes.

C.   1. c   2. a   3. b   4. b   5. a   6. c   7. b

D.   (1) Personal experience: this is the child's own past, present, and future. "When I was a baby."

     (2) Social activity: This is the child's routine. "After snack, it is music time."

     (3) Cultural time: This is time fixed by clocks and calendars. "The calendar tells us what day of the week it is."

E.   1. f   2. c   3. a, e, i   4. b, h   5. d, g

Unit 16 Graphs

A.   Graphs are used to show visually two or more comparisons in a clear way. The child when making a graph uses basic skills of classifying, comparing, counting, and measuring.

B.   (1) The child uses real objects to make his graph. Only two things are compared. One to one correspondence is used.

     (2) The child compares more than two items. A more permanent record is made.

     (3) The child uses more block charts. He starts to use cutout squares of paper. The child works more independently.

C.   1. c   2. a   3. b

D.   beads, interlocking cubes, spools, strings, paper squares.

Unit 17 Practical Activities

A.   1.   Each child can match, count, classify, compare, measure, and use relation concepts and number symbols as he plays house, store, and other grownup roles.

     2.   Children can learn the sequence from planting, growing, picking, buying, cooking, serving, and eating. They also count, measure, and match.

     3.   The teacher encourages dramatic role playing that helps children learn math by taking the children on field trips to farms, dairies, grocery stores, bakery shops, restaurants, the post office, or the bank. She also provides play props such as a cash register, purses, play money, and other things needed for various stores and shops.

     4.   The child can buy food.   The child can eat food.
          The child can prepare food.   (Also the child can grow and pick food.)
          The child can serve food.

B.   1. a   2. a/b   3. a   4. a/b   5. a   6. b

SECTION 4 SYMBOLS AND HIGHER LEVEL ACTIVITIES

Unit 18 Symbols

A.   1.   *Numerals* are the symbols used to represent amount.

     2.   The six number symbols skills are

        a. recognizing and saying the name of each numeral
        b. placing the numerals in order
        c. associating numerals with sets
        d. learning that each numeral in order stands for one more than the numeral that comes before it
        e. matching each numeral to any set of the size by which the numeral stands and making sets that match numerals
        f. writing numerals

     3.   *Self-correcting manipulative materials* are made so that the child can handle parts and solve the problem that the material poses.

     4.   The four basic types of self-correcting manipulative math materials are

        a. those which teach discrimination and matching
        b. those which teach sequence and order
        c. those which give practice in association of symbols with sets
        d. those which combine association of symbols and sets with sequence

     5.   The teacher's role is to show the children how to use the materials and then observe.

B.   1.   Recognition, sequencing, set association

     2.   Recognition, sequencing, associating numerals with sets

     3.   Recognition or sequencing

     4.   Recognition

     5.   Recognition and sequence

     6.   Recognition, sequencing, and set association

     7.   Recognition, sequence, and association

     8.   Recognition and sequence

     9.   Recognition, sequence, and association

    10.   Recognition, sequence

Unit 19 Sets and Symbols

A.	1.	higher level activity		5.	skill		9.	skill
	2.	skill		6.	skill		10.	higher level activity
	3.	higher level activity		7.	skill		11.	skill
	4.	skill		8.	skill		12.	skill

B.	1. c	2. a	3. b		
C.	1. b, c, d	2. c, d	3. a, b	4. a, b, c, d	5. b, c, d

Unit 20 Higher Level Activities

A. Five areas that include higher level activities are

classification, shapes, spatial relations, measurement, and graphs.

B. The three higher levels of classification are *multiple classification:* this is where the child classifies in more than one way such as size and shape.

class inclusion: this is where the child sees that one class may be part of another. For example, the child is given two oranges and three apples. When asked if there are more apples or more fruit, he says more fruit.

hierarchical classification: this is where the child sees that each item belongs to part of a larger category. For example, a shoe, a blouse, and mittens are all clothing items.

C.	1. c	2. a	3. b	4. a	5. c	6. b

D. 1. Mary shows an interest in spatial relations. The teacher could suggest that Mary draw a map of the street.

  2. Jim is classifying. The teacher can observe and then suggest that Jim look for different ways to classify the attribute blocks.

  3. Karen shows an interest in measurement. The teacher can bring over other objects to weigh. She can ask Karen how to make the scale balance.

  4. Alissa indicates an interest in shape. The teacher can ask her to find round objects in the room.

  5. The teacher can encourage the children to make a graph by coloring blue squares for those children who like blue best and yellow squares for those who choose the color yellow. This will show them whether blue or yellow is favored in the class.

SECTION 5 THE MATH ENVIRONMENT

Unit 21 Materials and Resources

A. 1. Manipulative math materials provide concrete experiences for children. They include blocks, balls, buttons, or anything the children use and handle.

  2. Pictorial math materials are semiconcrete. Children's picture storybooks are pictorial materials.

  3. Abstract math materials are symbolic. Dots and numerals are examples.

B. 1. In Teacher A's room there is not enough variety in the selection of materials. The geoboard can not be used without the rubber bands. Teacher B has a better selection as there is more variety.

  2. Yes, she is teaching math. Maybe Teacher C thinks teaching means telling. However, she has a responsive math environment as she has provided concrete materials for the children to learn. Tell her to watch the children playing in the sand with the measuring materials.

  3. First, tell her to take an inventory of what she already has. Then, ask her to categorize these into the math skill areas — matching, counting, and so on. She will then be able to determine what she needs. Next, tell her to shop and check prices at local stores or look through catalogs. She can now decide if she wants to buy all the materials, or if she wants to make some or ask for donations.

4. The teacher should first look at where the math materials are placed in the environment and observe how the children are using the materials. This information will help her decide what to do next.

C. 1. This center needs more classifying items, shape materials, and pattern materials. Possibly some "money" items would also be helpful.

2. Kindergarten children still need concrete materials as well as abstract. A selection of manipulative math materials should be available for the children to use.

D.
1. i	4. a	7. f	10. d, e	13. d
2. h	5. b	8. i	11. a	14. g
3. f	6. d, e	9. i	12. c	15. f

Unit 22 Math in Action

A.
1. b	3. c	5. b	7. a	9. b
2. d	4. d	6. b	8. d	10. c

B. 1. Buy proper equipment. Have odds and ends available.

2. Buy quality blocks and a good quantity. Make blocks accessible on low open shelves.

3. Games that teach math include board games such as Candyland, aiming such as bowling, or action games such as musical chairs. Various games should be available. These require teacher guidance. Competition among children should not be encouraged.

4. The teacher should know a few math finger plays and songs. She repeats the words to the children. She demonstrates the motions. Together they act out the finger play or song.

Unit 23 Math in the Home

A. The first parent statement is positive and would get the child interested in the activity.

The second parent statement is negative. This is a put down and would stifle the child's interest in doing the task.

The parents "Yeah" is not very supportive.

The parent could have used more praise before asking the child about the smaller items.

The last statement could be positive or negative. The tone of voice would affect the intent. It would be better to say, "I like the way you can find bigger and smaller objects. Let's see if we can find longer and shorter things in the house."

B. 1. There are many reasons why teachers need to learn how to work with parents. These include recognizing that parents are the children's first and most important teacher. Parents often ask teachers for advice on how to work with their children. They ask what toys to buy for their children.

2. Approaches include newsletters, open houses, and conferences.

3. Guides for helping parents become teachers include patience, praise, repetition, and the use of concrete experiences.

4. Naturalistic activities include counting the buttons on the child's sweaters and saying such things as, "After breakfast it will be 9:00 o'clock."

Informal activities include having the child fold napkins or weighing the child.

Structured activities include making cardboard shapes and using them for activities or making look alike cards and having the child match sets.

Acknowledgments

The authors wish to express their appreciation to the following individuals and Early Childhood Education and Development Centers:

- Jerry Bergman for many of the photographs

- Roger, Brent, and Dean Radeloff and Jim and Kate Charlesworth for their patience, encouragement, and active participation in this endeavor

- The children and teachers who were photographed in the listed Early Childhood Centers:

  Plan-Do-Talk Day Care Center, Bowling Green, Ohio
  Dunn's Kiddie Kare, Bowling Green, Ohio
  Bowling Green State University Nursery School, Bowling Green, Ohio
  Pineview Preschool, Albany, New York

- Our other friends and relatives who served as photographic subjects

- The three talented ladies who typed the final manuscript: Mrs. Sherry Haskins, Mrs. Doris Taube, and Mrs. Elaine Thomas

- The staff at Delmar Publishers, with special thanks to Elinor Gunnerson and Alan Knofla for their faith in unknown authors

Delmar Staff

Director of Publications — Alan N. Knofla
Source Editor — Elinor Gunnerson
Consultant — Jeanne Machado
Photographer — Joseph Tardi Associates

Sections of this edition were classroom tested at Bowling Green State University, Ohio

INDEX

10/86 (5C1851D)